WORKBOOK / STUDY GUIDE TO ACCOMPANY

METEOROLOGY TODAY

AN INTRODUCTION TO WEATHER, CLIMATE, AND THE ENVIRONMENT

FOURTH EDITION

C. DONALD AHRENS

WEST PUBLISHING COMPANY

St. Paul New York Los Angeles San Francisco

50 W. Kellogg Boulevard
P.O. Box 64526
St. Paul, MN 55164-1003

Printed in the United States of America
99 98 97 96 95 94 93 92 8 7 6 5 4 3 2 1 0

ISBN 0–314–86294–3

CONTENTS

PREFACE

This workbook/study guide has been developed as a tool to help you learn the information in the textbook, *Meteorology Today,* 4th Edition. Meteorology is a complex subject; understanding weather and climate means mastering many new terms, concepts and ideas. It is my hope that with the aid of this workbook/study guide, learning about the weather around you will be a positive and enjoyable experience.

The workbook/study guide is organized so that each chapter begins with a brief *summary* of the information presented in the corresponding textbook chapter. This is followed by a list of the chapter's *important concepts.* Continuing through the workbook/study guide you will find *matching, fill-in-the-blank, multiple choice, true-false* and *problem type* questions. I feel the best way to answer these questions is to first carefully read the chapter in your textbook, then begin answering the questions in the study guide. If you find that you are having difficulty with a certain concept, return to your textbook and reread the section dealing with that particular topic. When answering the true-false questions it is important that you know why an answer is false, not just that it is. After you have finished answering the questions, turn to the end of the workbook/study guide chapter for the correct answers.

Toward the end of each chapter is a list of *additional readings.* They provide a closer look at certain issues and subjects associated with the topics covered in the chapter. Most are found in journals that are available in university and college libraries. I have also listed a few books that may be helpful.

The study of the science of meteorology is fascinating and often exciting. It is my hope that through your text, this study guide, and the information you gain in class you will come to a special appreciation of the dynamics of the atmosphere around you.

ACKNOWLEDGMENTS

Special thanks to the many people who helped with the preparation of this workbook/study guide: To Joe Medeiros for his talented and creative illustrations; to Laurie Rydelius, Nancy Newsom and Kay Behrens for carefully reading the manuscript; to Martin Arthur for his layout and production; and to Suzy Ahrens whose input into this project has been invaluable. A special note of gratitude goes to Steve Schonebaum for his professional concern, editing, and assistance.

CHAPTER

1

INTRODUCTION: THE EARTH AND ITS ATMOSPHERE

Chapter One is a rather brief introductory chapter that provides you with a broad overview of the earth and its atmosphere. The first part of the chapter focuses on the earth as a planet. Here we see that the earth is one of nine planets, with a unique atmosphere comprised of nitrogen and oxygen with small amounts of other gases. Then comes an overview of the earth's atmosphere. Here we get a glimpse of several topics that will be covered more completely in later chapters, such as storm systems, weather maps, and satellite photographs. Keep in mind that the material in this section is mainly introductory in nature. It is designed to familiarize you with the atmosphere and some of the processes that take place within it. The last section briefly details the many ways weather and climate influence our lives, from how we feel on a windy day to the type of clothing we should purchase for the coming season.

Some important concepts of this chapter:

1. The atmosphere of Earth is mainly nitrogen and oxygen, while the atmospheres of its closest neighbors, Venus and Mars, are mainly carbon dioxide.

2. The earth's atmosphere contains storms of all sizes, ranging from huge middle latitude cyclonic storms that may extend for thousands of kilometers, to the much smaller tornado that is typically less than one kilometer wide.

3. The wind direction is the direction *from which* the wind is blowing. A north wind blows *from* the north.

4. In the Northern Hemisphere, at the surface the wind blows clockwise and outward around an area of high pressure, and counterclockwise and inward around an area of low pressure.

5. Weather is the condition of the atmosphere at any particular time and place. Climate is the accumulation of daily and seasonal weather events that occur over a given period of time.

SELF-TESTS

MATCH THE FOLLOWING

______ 1. Another name for a high pressure area

______ 2. The planet with a strong greenhouse effect, whose surface temperature averages 480°C (900°F)

______ 3. A storm of tropical origin with winds in excess of 64 knots (74 mi/hr)

______ 4. An elongated region of low atmospheric pressure

______ 5. Planet whose atmosphere is mainly nitrogen and oxygen

______ 6. Another name for an extratropical cyclone

______ 7. A towering cloud (or cluster of clouds) accompanied by thunder, lightning, and strong gusty winds

______ 8. A relatively small, rotating funnel that extends downward from the base of a thunderstorm

______ 9. The study of the atmosphere and its phenomena

______ 10. On a weather map, this zone marks sharp changes in temperature, humidity, and wind direction.

______ 11. It holds a planet's atmosphere close to its surface

______ 12. The horizontal movement of air

a. hurricane

b. wind

c. thunderstorm

d. Earth

e. tornado

f. gravity

g. anticyclone

h. meteorology

i. trough

j. Venus

k. front

l. middle latitude cyclonic storm

FILL IN THE BLANK

1. The primary source of energy for the earth's atmosphere is the __________.

2. The gas primarily responsible for the greenhouse effect on Venus is __________________.

3. Where cold surface air is replacing warm surface air, the boundary separating the different bodies of air is termed a __________ __________.

4. A molecule is a combination of two or more __________.

5. An elongated region of high pressure is called a __________.

6. Where warm surface air is replacing cold surface air, the boundary separating the different bodies of air is called a __________ __________.

7. Horizontal differences in air pressure produce __________.

MULTIPLE CHOICE

1. The largest storm in our atmosphere, in terms of actual size (diameter) is the:

   a. thunderstorm
   b. hurricane
   c. middle latitude cyclonic storm
   d. tornado

2. If you are standing north of a smoke stack and smoke from the stack is drifting over your head, the wind direction would be:

   a. north   b. south

3. In the Northern Hemisphere, surface winds tend to blow this way around an area of *surface low pressure.*

   a. clockwise and inward
   b. clockwise and outward
   c. counterclockwise and inward
   d. counterclockwise and outward

4. Which of the following statements relates to weather rather than climate?

   a. Outside it is sunny and hot.
   b. The winters here are cold and wet.
   c. Thunderstorms are prevalent during July.
   d. Our lowest temperature ever, was -20°C (-4°F).
   e. The average temperature during March is 15°C (59°F).

5. Of these four storms, the smallest in terms of actual size (diameter) is:

   a. thunderstorm
   b. hurricane
   c. middle latitude cyclonic storm
   d. tornado

6. Which below is *not* considered a weather element?

   a. air pressure
   b. air temperature
   c. clouds
   d. wind
   e. air density

7. In the adjacent weather map, a cold front is located between numbers _____ and _____ while a warm front is between numbers _____ and _____.

   a. 1 and 2; 2 and 4
   b. 1 and 2; 2 and 3
   c. 2 and 3; 3 and 5
   d. 2 and 3; 2 and 4
   e. 2 and 4; 1 and 2

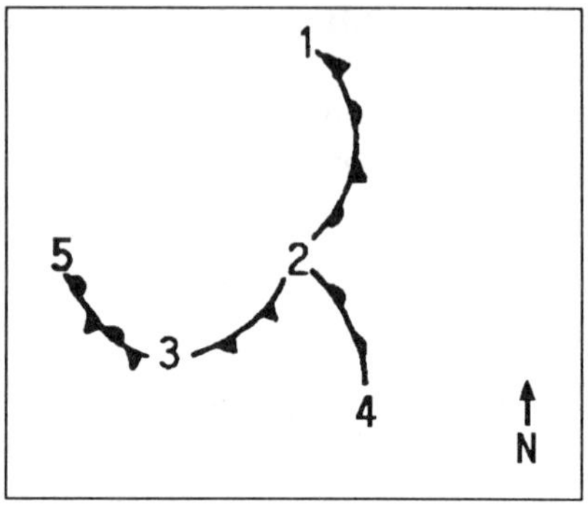

Surface Weather Map

■ TRUE - FALSE

_____ 1. A north wind blows from the north and a south wind blows from the south.

_____ 2. The most abundant gas in the earth's atmosphere by volume is carbon dioxide.

_____ 3. In the Northern Hemisphere, surface winds tend to blow clockwise and outward around an area of high pressure.

_____ 4. Generally, weather in the middle latitudes moves from east to west.

_____ 5. Normally, large areas of high pressure tend to be accompanied by fair weather.

ADDITIONAL READINGS

"Saturn: Weather Under the Rings" by Alan Dyer, *Weatherwise*, Vol. 44, No. 3 (June 1991), p. 19.

"Venus: Weather like Hell" by Alan Dyer, *Weatherwise*, Vol. 44, No. 2 (April 1991), p. 10.

"From Gods to Satellites" by John Farrand, Jr., *Weatherwise*, Vol. 44, No. 2 (April 1991), p. 30.

"To Understand the Atmosphere" by Henry Lansford, *Weatherwise*, Vol. 38, No. 4 (August 1985), p. 184.

"Weather, Climate and You" by H. E. Landsberg, *Weatherwise*, Vol. 39, No. 5 (October 1986), p. 248.

"The Climate of Mars" by Robert M. Haberle, *Scientific American*, Vol. 254, No. 5 (May 1986), p. 54.

"Weather, Climate and the Economy" by Patrick Hughes, *Weatherwise*, Vol. 35, No. 2 (April 1982), p. 60.

ANSWERS

MATCHING

1. g
2. j
3. a
4. i
5. d
6. l
7. c
8. e
9. h
10. k
11. f
12. b

FILL IN THE BLANKS

1. sun
2. carbon dioxide (CO_2)
3. cold front
4. atoms
5. ridge
6. warm front
7. wind

MULTIPLE CHOICE

1. c
2. b
3. c
4. a
5. d
6. e
7. d

TRUE - FALSE

1. T
2. F
3. T
4. F
5. T

CHAPTER

2

THE EARTH'S ATMOSPHERE: ORIGIN, COMPOSITION, AND STRUCTURE

Chapter Two describes the earth's atmosphere in more detail. It begins by presenting a possible theory as to how the earth's atmosphere may have evolved. This discussion leads to the examination of the various constituents found in today's atmosphere, including the different greenhouse gases and the major atmospheric pollutants. The next section describes the different phases of water, as well as how water is transported and circulated within our atmosphere. This is followed by a section that investigates the behavior of air. Here we consider important relationships between air temperature, pressure, and density. We also examine temperature scales and learn how barometers work. The last section describes how atmospheric temperature varies with height and how this temperature profile can be divided into layers. Here we see that stratospheric ozone is being depleted by a number of complex chemical reactions and that an ozone hole has actually formed over the continent of Antarctica.

Some important concepts of this chapter:

1. In a volume of air near the earth's surface, nitrogen occupies about 78 percent and oxygen about 21 percent.
2. The primary greenhouse gases in the earth's atmosphere include: water vapor (H_2O), carbon dioxide (CO_2), methane (CH_4), nitrous oxide (N_2O), and chlorofluorocarbons (CFCs).
3. Water is the only substance in our atmosphere that is found naturally as a liquid (water), as a gas (water vapor), and as a solid (ice).
4. The temperature of a substance is a measure of the average speed of its molecules.
5. Atmospheric pressure is a measure of the total weight of air above any point.
6. At the same temperature and at the same level, warm humid air is less dense than warm dry air.
7. Rising air expands and cools, while sinking air is compressed and warms.
8. At the same level in the atmosphere (same air pressure), a parcel of cold air will be more dense than a parcel of warm air.
9. A measured increase in air temperature with height is called an *inversion*, while the rate at which the air temperature decreases with height is called the *lapse rate*.
10. The majority of ozone in our atmosphere is found in the stratosphere, the layer above the troposphere.

SELF TESTS

MATCH THE FOLLOWING

______ 1. Weather element that *always* decreases with increasing height

______ 2. The change of state of water vapor into a liquid

______ 3. The process which causes rising air to cool

______ 4. A barometer that contains no liquid

______ 5. Instrument that measures temperature, pressure and humidity at levels above the earth's surface

______ 6. The most abundant greenhouse gas in the earth's atmosphere

______ 7. The unit of pressure most commonly found on a surface weather map

______ 8. Gas that strongly absorbs ultraviolet radiation in the stratosphere

______ 9. A measure of the average speed of air molecules

______ 10. A condition caused by a lack of oxygen to the brain

______ 11. The outpouring of gases from the earth's hot interior

______ 12. Greenhouse gas used as a refrigerant and as a solvent

______ 13. The change of state of ice into vapor

______ 14. A temperature scale where 0° represents freezing and 100° boiling

______ 15. Layer of the atmosphere that contains almost all the weather

______ 16. Temperature scale that begins at absolute zero

______ 17. Instrument that measures pressure and indicates altitude

______ 18. The average decrease in air temperature with increasing height above the surface

a. temperature

b. expansion

c. ozone

d. altimeter

e. condensation

f. troposphere

g. outgassing

h. Celsius

i. radiosonde

j. pressure

k. millibar

l. aneroid

m. chlorofluorocarbons

n. hypoxia

o. sublimation

p. lapse rate

q. water vapor

r. Kelvin

FILL IN THE BLANK

1. The gas that shows the most variation from place to place and from time to time in the lower atmosphere is ____________.

2. The mass of air in a given volume describes the air's ________.

3. Most of the ozone in the atmosphere is found in the atmospheric layer called the ________.

4. The circulation of water within the atmosphere is called the ________ cycle.

5. What percent does each of the following gases occupy in a volume of air near the earth's surface?
   nitrogen _____%
   oxygen _____%
   water vapor _____%
   carbon dioxide _____%

6. A recording aneroid barometer can also be called a ________.

7. Thermal (atmospheric) tides cause a daily (diurnal) rise and fall in what weather element at the earth's surface? ________

8. When chlorofluorocarbons (CFCs) are subjected to ultraviolet radiation from the sun, ________ is released, a gas which rapidly destroys ozone.

9. Energy of motion is also known as ________ ________.

10. Most of the earth's water is believed to have originally come from ____________.

11. A temperature of 300 K is equal to ________ degrees Celsius.

12. The process of water changing from a liquid to a vapor is called ________.

13. An invisible balloon-like body of air is called a ________.

14. The hottest atmospheric layer is the ________.

15. The only substance near the earth's surface that is found naturally in the atmosphere as a solid, a liquid, and a gas is ________.

16. At sea level, the average or standard value of atmospheric pressure is ________ millibars and ________ inches of mercury.

17. A measured increase in air temperature with increasing height is called an __________.

18. The __________ temperature scale sets the freezing of pure water at 32°.

19. The electrified region of the upper atmosphere is called the __________.

20. The average decrease in air temperature with increasing height in the lower atmosphere is about ______°C per 1000 meters, or ______°F per 1000 feet.

21. The atmospheric layer in which we live is called the __________.

22. The primary ingredient of photochemical smog is __________.

MULTIPLE CHOICE

1. Which below is *not* a process by which CO_2 enters the atmosphere?

   a. volcanic eruptions
   b. decay of vegetation
   c. evaporation
   d. burning of fossil fuels

2. The process by which water changes from a vapor to a solid is called:

   a. evaporation
   b. condensation
   c. transpiration
   d. deposition

3. The *primary* source of oxygen for the earth's atmosphere during the past half billion years or so appears to be:

   a. volcanic eruptions
   b. photosynthesis
   c. photodissociation
   d. exhalation of animal life
   e. transpiration

4. The "ozone hole" is observed in this atmospheric layer:

   a. thermosphere
   b. mesosphere
   c. troposphere
   d. stratosphere
   e. ionosphere

5. Air density normally:

   a. increases with increasing height
   b. decreases with increasing height
   c. remains constant with increasing height

6. In the thermosphere the temperature can exceed 1000°C (1800°F), yet a person's exposed hand might not "feel" hot, primarily because:

   a. the molecules are not moving rapidly at that elevation
   b. there is no energy associated with the air molecules in the upper atmosphere
   c. the air density of the atmosphere is low, so few molecules would collide with a person's hand
   d. the air pressure is extremely low in the thermosphere
   e. you lose all sense of feeling at those elevations

7. The so-called "ozone hole" is observed above:

   a. the continent of North America
   b. the continent of Asia
   c. the equator
   d. the continent of Australia
   e. the continent of Antarctica

8. Which below is *not* considered a greenhouse gas?

   a. carbon dioxide (CO_2)
   b. methane (CH_4)
   c. water vapor (H_2O)
   d. nitrous oxide (N_2O)
   e. oxygen (O_2)

9. A force exerted on a unit area describes air __________.

   a. density
   b. temperature
   c. pressure

10. As the average speed of air molecules decreases, the temperature of the air __________.

   a. increases
   b. decreases
   c. does not change

11. When the surface air pressure increases, the height of a column of mercury in a barometer will:

    a. also increase
    b. decrease
    c. show no change

12. Which of the following relationships best describes the gas law?

    a. pressure is proportional to density times temperature
    b. temperature is proportional to density times pressure
    c. density is proportional to pressure times temperature

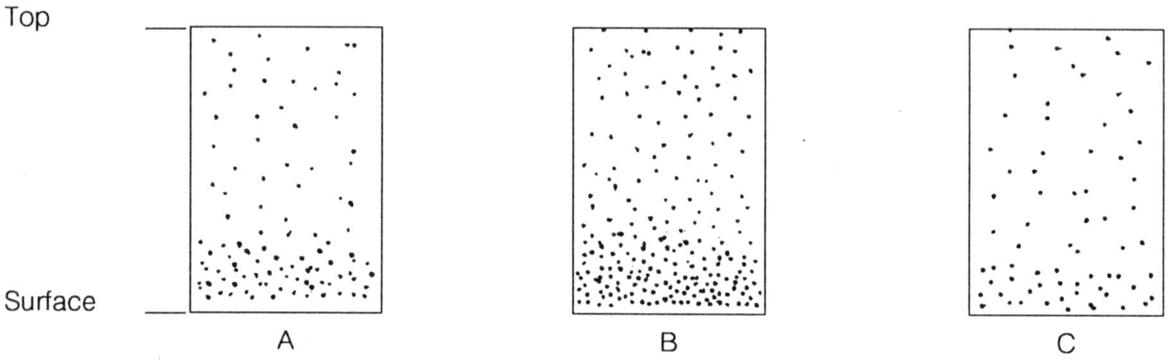

Figure 1

13. In Figure 1, if the air temperature is the same in each column, which column has the highest surface air pressure? (Each dot represents billions of air molecules.)

    a. column A
    b. column B
    c. column C

■ TRUE - FALSE

______ 1. Over the last 100 years, the concentration of CO_2 in the earth's atmosphere has been increasing.

______ 2. Chlorofluorocarbons enter the atmosphere mainly through the decay of vegetation.

______ 3. All other factors being equal, increasing wind speed enhances evaporation.

______ 4. A degree Fahrenheit is larger than a degree Celsius.

______ 5. Near the earth's surface at the same temperature and level in the atmosphere, warm humid air is less dense than warm dry air.

_____ 6. An air temperature of 0°K would be the same as an air temperature of 0°C.

_____ 7. On a clear, sunny day an aneroid barometer carried to the top of a mountain would indicate clear, sunny weather.

_____ 8. At a constant pressure, cold air is more dense than warm air.

_____ 9. The atmosphere is compressible. This means that air pressure decreases at a constant rate from the surface to the top of the atmosphere.

_____ 10. In Los Angeles, California, the word "smog" means the combination of smoke and fog.

_____ 11. Water vapor is a gas.

_____ 12. The atmospheric concentration of carbon dioxide (CO_2) is generally higher in summer than in winter.

_____ 13. The air temperature at the tropopause is much colder than the average air temperature measured at the earth's surface.

_____ 14. As the air temperature increases, condensation becomes *less* likely.

_____ 15. The two most abundant gases in the stratosphere are nitrogen and oxygen.

_____ 16. The highest level of the ionosphere is called the D region.

_____ 17. In the middle latitudes a pressure of 500 millibars would normally be found at an altitude near 5500 meters or about 18,000 feet.

_____ 18. Sinking air always warms and rising air always cools.

_____ 19. In the upper atmosphere, the mean free path of atoms and molecules decreases with increasing altitude.

_____ 20. A change in air density above an area can bring about a change in surface air pressure.

■ PROBLEMS

1. a. Use the gas law to calculate the air pressure in millibars when the air temperature is -56°C, and the air density is 0.161kg/m^3. (Hint: use the gas law in the Focus Section on p. 38 of your text.)

   b. Look at Fig. 2.16 (p. 39 in textbook) and determine at approximately what altitude (in miles) you would expect to observe the altitude you calculated in (a). Since 1 mi = 1.61 km, this altitude would be equal to how many kilometers?

   c. What atmospheric layer would you be in? __________

   What percent of all the air molecules would be below you? __________

2. If one inch of mercury equals 33.865 millibars, convert the following readings to millibars.

   a. 31.0 inches __________ millibars

   b. 30.5 inches __________ millibars

   c. 30.0 inches __________ millibars

   d. 29.5 inches __________ millibars

   e. 29.0 inches __________ millibars

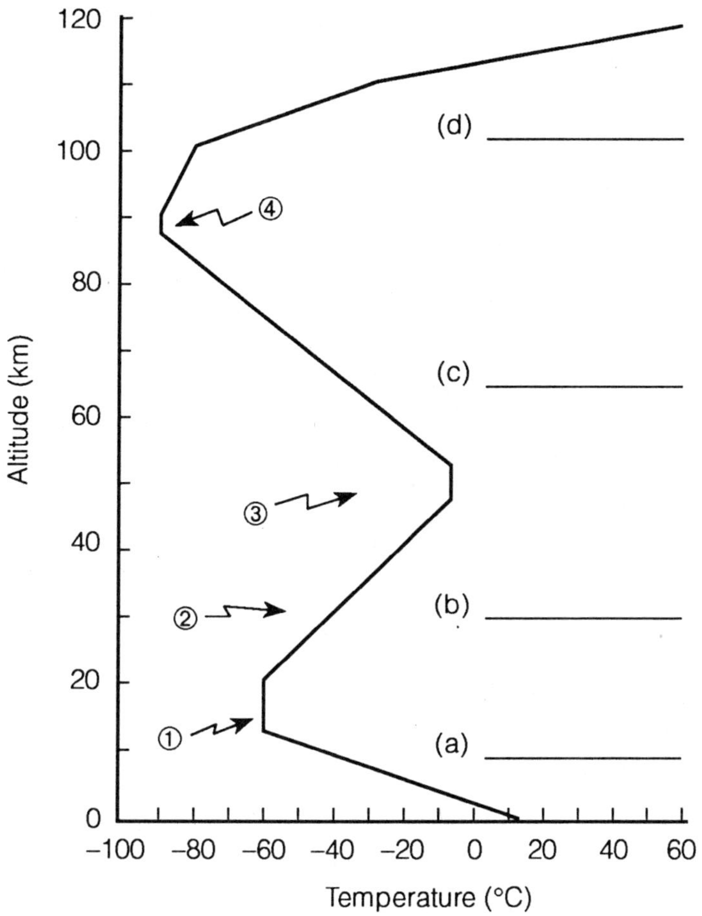

Figure 2

3. The following questions refer to Figure 2.

   a. In the diagram label the four layers of the atmosphere in the blank spaces provided.

   b. The maximum concentration of ozone would be observed at number _____.

   c. Which number is pointing to the stratopause?

      a. 1 b. 2 c. 3 d. 4

   d. Which number is pointing to the tropopause?

      a. 1 b. 2 c. 3 d. 4

4. In the spaces provided in Figure 3, convert the temperature from degrees Fahrenheit into degrees Celsius. (Hint: conversion factors are found in appendix A, p. 522 of your textbook.)

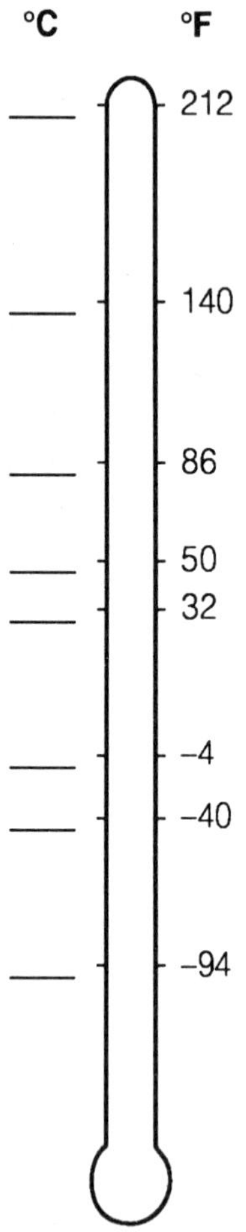

Figure 3

■ ADDITIONAL READINGS

"The Global Carbon Cycle" by Wilfred M. Post, et. al., *American Scientist*, Vol. 78, No. 4 (July August 1990), p. 310.

"How Bleak is the Outlook for Ozone?" by W. Hively, *American Scientist*, Vol. 77, No. 3 (May-June 1989), p. 219.

"Chlorofluorocarbons and the Depletion of Stratospheric Ozone" by F. Sherwood Rowland, *American Scientist,* Vol. 77, No. 1 (Jan-Feb 1989), p. 36.

"Agriculture and Carbon Dioxide" by Paul E. Waggoner, *American Scientist,* Vol, 72, No. 2 (March-April 1984), p. 179.

"The Antarctic Ozone Hole" by Richard S. Stolarski, *American Scientist,* Vol. 258, No. 1 (January 1988), p. 30.

"Ozone Crises" by William Brune, *Weatherwise,* Vol. 43, No. 3 (June 1990), p. 136.

"A Matter of Degrees" by Kirby Roche, *Weatherwise,* Vol. 43, No. 4 (August 1990), p. 200.

"An Inventory of the World's Water" by Frank H. Forrester, *Weatherwise,* Vol. 38, No. 2 (April 1985), p. 83.

"Polar Stratospheric Clouds and Ozone Depletion" by Owen B. Toon and Richard P. Turco, Vol. 264, No. 6 (June 1991), p. 68.

"How Climate Evolved on the Terrestrial Planets" by James F. Kasting, Owen B. Toon, and James B. Pollack, *Scientific American,* Vol. 254, No. 5 (May 1986), p. 54.

"Atmospheric Trace Gases: Trends and Distributions over the Last Decade" by R. A. Rasmussen and M. A. K. Khalil, *Science,* Vol. 232, No. 4758 (June 27, 1986), p. 1623.

"Conceptions and Misconceptions of Pressure" by Craig F. Bohren, *Weatherwise,* Vol. 36, No. 2 (April 1983), p. 82.

"Carbon Dioxide" by Kirby Hanson, *Weatherwise,* Vol. 33, No. 6 (December 1980), p. 253.

ANSWERS

MATCHING

1. j
2. e
3. b
4. l
5. i
6. q
7. k
8. c
9. a
10. n
11. g
12. m
13. o
14. h
15. f
16. r
17. d
18. p

FILL IN THE BLANK

1. water vapor
2. density
3. stratosphere
4. hydrologic
5. nitrogen = 78%
   oxygen = 21%
   water vapor = 0 to 4%
   carbon dioxide = 0.035%
6. barograph
7. pressure
8. chlorine
9. kinetic energy
10. inside the earth
11. 27°C
12. evaporation
13. parcel
14. thermosphere
15. water
16. 1013.2 mb, 29.92 in.
17. inversion
18. Fahrenheit
19. ionosphere
20. 6.5°C/1000 m, 3.6°F/1000 ft
21. troposphere
22. ozone

MULTIPLE CHOICE

1. c
2. d
3. b
4. d
5. b
6. c
7. e
8. e
9. c
10. b
11. a
12. a
13. b

TRUE - FALSE

1. T
2. F
3. T
4. F
5. T
6. F
7. F
8. T
9. F
10. F
11. T
12. F
13. T
14. T
15. T
16. F
17. T
18. T
19. F
20. T

PROBLEMS

1. a. ~ 100 mb
   b. ~ 10 mi, 16.1 km
   c. stratosphere, about 90 percent
2. a. 1049.8 mb
   b. 1032.9 mb
   c. 1015.9 mb
   d. 999.0 mb
   e. 982.1 mb
3. a. (a) troposphere
      (b) stratosphere
      (c) mesosphere
      (d) thermosphere
   b. number 2
   c. number 3
   d. number 1
4. -94°F = -70°C
   -40°F = -40°C
   -4°F = -20°C
   32°F = 0°C
   50°F = 10°C
   80°F = 30°C
   140°F = 60°C

CHAPTER

3

WARMING THE EARTH AND THE ATMOSPHERE

Chapter Three examines the concepts of energy, heat, and heat transfer. First it looks at the various forms of energy and how energy in our atmosphere is transferred from one form into another. Next, it considers the topics of absorption and emission of energy in terms of heating and cooling the earth. At this point we examine the atmospheric greenhouse effect and the various gases that produce it. Here we learn that without a greenhouse effect, the average surface temperature of our planet would be considerably colder than it is now. We also see that it is the enhancement of the greenhouse effect by the increasing concentration of greenhouse gases that is of concern to most scientists. The section that follows describes how the earth and its atmosphere are warmed. Included here is a section on the energy balance of the earth and its atmosphere. The final section looks closely at the sun's energy and its influence on our atmosphere.

Some important concepts of this chapter:

1. Evaporation is a cooling process and condensation is a warming process.
2. The transfer of heat within our atmosphere can take place by conduction, convection and radiation.
3. All objects with a temperature above absolute zero, -273°C (-459°F), emit radiation.
4. The higher an object's temperature, the more total radiation emitted each second by the object, and the shorter the wavelength of maximum emission.
5. The earth's surface absorbs solar radiation only during the daylight hours; however, it constantly emits infrared radiation, both during the day and at night.
6. Water vapor and carbon dioxide are important atmospheric greenhouse gases that selectively absorb and emit infrared radiation, thereby keeping the earth's average surface temperature warmer than it would be otherwise.
7. Enhancement of the atmospheric greenhouse effect may be taking place because of the increasing concentrations of greenhouse gases.
8. The lower atmosphere is mainly heated from below.
9. The annual average temperature of the earth and the atmosphere remains fairly constant from one year to the next because the amount of energy they absorb each year is equal to the amount of energy they lose.
10. The aurora is produced in the upper atmosphere as energetic particles collide with atoms and molecules, which become excited and emit energy in the form of visible light.

SELF TESTS

MATCH THE FOLLOWING

______ 1. Heat transfer process that depends upon the movement of air	a.	conduction
______ 2. Energy of motion	b.	sensible heat
______ 3. Rising bubbles of air	c.	infrared
______ 4. The heat we can feel and measure with a thermometer	d.	kinetic energy
______ 5. The horizontal transfer of any atmospheric property by the wind	e.	specific heat
______ 6. Energy transferred by electromagnetic waves	f.	solar wind
______ 7. One millionth of a meter	g.	convection
______ 8. Charged particles traveling through space	h.	aurora
______ 9. The horizontal distance between two wave crests	i.	radiation
______ 10. This is released as sensible heat during the formation of clouds	j.	wavelength
______ 11. The transfer of heat by molecule-to-molecule contact	k.	visible
______ 12. The sun emits radiation with greatest intensity in this region of the spectrum	l.	thermals
______ 13. The heat capacity of a unit mass of a substance	m.	latent heat
______ 14. Wavelengths longer than those of red light	n.	micrometer
______ 15. Visible light given off by excited atoms and molecules in the upper atmosphere	o.	ultraviolet
______ 16. Electromagnetic waves whose wavelengths are shorter than those of visible light	p.	advection

▩ FILL IN THE BLANK

1. Sunlight that bounces off a surface is said to be _________.

2. A perfect absorber and a perfect emitter of radiation is called a _________ _________.

3. How much radiation would an object be emitting if its temperature is at absolute zero? _________

4. The wavelength range where neither water vapor nor carbon dioxide absorb much of the earth's infrared radiation is known as the atmospheric _________.

5. The _________ represents the reflectivity of a surface.

6. The two most significant atmospheric greenhouse gases in the earth's atmosphere are _________ and _________.

7. In the Northern Hemisphere another name for the northern lights is the _________ _________.

8. Cooler, dark blemishes on the sun are called _________.

9. At night objects on the ground cool by the process of emitting _________ _________.

10. The combined albedo of the earth and its atmosphere averages about _________ percent.

11. The sun emits maximum radiation at visible wavelengths near _________ micrometers, while the earth emits maximum radiation of infrared wavelengths near _________ micrometers.

12. The temperature at which the earth is both absorbing solar radiation and emitting infrared radiation at equal rates is called the Earth's _________ _________ _________.

13. If the present concentration of CO_2 doubles in 100 years, climate models predict that for the earth's average temperature to rise 5°C, the gas _________ _________ must also increase in concentration.

■ MULTIPLE CHOICE

1. The proper order of waves from longest to shortest is:

   a. visible, infrared, ultraviolet
   b. infrared, visible, ultraviolet
   c. ultraviolet, visible, infrared
   d. visible, ultraviolet, infrared
   e. ultraviolet, infrared, visible

2. Heat is energy in the process of being transferred from:

   a. low pressure to high pressure
   b. cold objects to hot objects
   c. high pressure to low pressure
   d. hot objects to cold objects
   e. regions of low density toward regions of high density

3. The rate at which energy is emitted by a body:

   a. increases with decreasing temperature
   b. increases with increasing temperature
   c. does not depend on the body's temperature

4. If the earth's average surface temperature should decrease, the amount of radiation emitted from the earth's surface will __________, and the wavelength of maximum emission will shift toward __________ wavelengths.

   a. increase, shorter
   b. increase, longer
   c. decrease, shorter
   d. decrease, longer

5. If the earth had no atmospheric greenhouse effect, the average surface temperature would be:

   a. lower than at present
   b. higher than at present
   c. the same as it is now

6. The albedo of the moon is 7 percent. This means that:

   a. 7 percent of the sunlight that strikes the moon is reflected
   b. 7 percent of the sunlight that strikes the moon is absorbed
   c. 93 percent of the sunlight that strikes the moon is reflected
   d. 93 percent of the sunlight that strikes the moon is conducted toward its center
   e. 7 percent of the sunlight that strikes the moon is transferred upward by convection

7. Clear, calm nights are usually cooler than cloudy, calm nights because:

   a. clear skies always contain less water vapor than cloudy skies
   b. the clouds start convection currents between them
   c. cloud droplets are better conductors of heat than is a clear sky
   d. clouds absorb and radiate infrared radiation back to the earth's surface
   e. water droplets in clouds reflect infrared energy back to the earth's surface

8. The moon's surface can only cool by (Hint: the moon has no atmosphere):

   a. convection
   b. conduction
   c. radiation

9. The earth's atmospheric greenhouse effect is produced mainly by water vapor and carbon dioxide absorbing and re-emitting:

   a. visible radiation
   b. infrared radiation
   c. ultraviolet radiation

10. Which of the gases below is *not* believed responsible for enhancing the earth's atmospheric greenhouse effect?

   a. molecular oxygen (O_2)
   b. carbon dioxide (CO_2)
   c. methane (CH_4)
   d. nitrous oxide (N_2O)
   e. chlorofluorocarbons (CFCs)

TRUE - FALSE

______ 1. The sun's radiation is also referred to as shortwave radiation.

______ 2. Clouds are poor absorbers and emitters of infrared radiation.

______ 3. Only selective absorbers in the atmosphere emit radiation.

______ 4. The term "latent" means hidden.

______ 5. On the average, about 50 percent of the solar radiation that strikes the outer atmosphere eventually reaches the earth's surface.

______ 6. An ultraviolet photon carries less energy than an infrared photon.

______ 7. Air is a poor conductor of heat.

______ 8. The earth's atmosphere behaves as a black body.

______ 9. On the average, each year the earth-atmosphere system sends off into space just as much energy as it receives.

______ 10. The earth emits most of its radiation in the ultraviolet portion of the spectrum.

______ 11. On the average, 75 percent of the sunlight that strikes water daily is reflected.

______ 12. Energy is the ability or capacity to do work.

______ 13. An increase in cloud cover around the earth would probably increase the albedo of the earth-atmosphere system.

______ 14. Air glow is another name for the aurora.

______ 15. The processes of condensation, freezing, and deposition all release sensible heat into the environment.

■ PROBLEMS

1. a. Suppose the sun's average surface temperature increases from 6000 K to 12,000 K. Use Wien's Law on p. 59 of your textbook to calculate the wavelength (in micrometers) at which the sun would be emitting the majority of its radiation.

   b. This wavelength (see a. above) is in what region of the electromagnetic spectrum? __________

   c. By doubling the absolute temperature of the sun's surface, the sun's energy output would increase by a factor of __________.

Figure 1

2. Figure 1 shows a calm day with clouds (a), a calm day without clouds (a'), a calm night with clouds (b), and a calm night without clouds (b').

   a. If both surfaces above a and a' are covered with grass (and all other factors are the same), would the surface air temperature most likely be higher in a or a'? __________

   b. Would the albedo of the earth-atmosphere system most likely be greater in a or a'? __________

   c. If both surfaces above b and b' are bare soil (and all other factors are the same), would the surface air temperature most likely be higher in b or b'? __________

d. Suppose of the four diagrams (a, a', b, b'), the highest surface air temperature is observed in a'. Then, in which of the four conditions (a, a', b, b') would the earth's surface be emitting the most infrared radiation? __________

e. If you answered problem (d) correctly, explain in terms of absorption and emission of radiation why the highest (and not the lowest) temperature is observed at a'.

■ ADDITIONAL READINGS

"The Aurora" by Syun-Ichi Akasofu, *American Scientist*, Vol. 69, No. 5 (Sept-Oct 1981), p. 492.

"The Awesome Aurora" by Roy S. Hall, *Weatherwise*, Vol. 40, No. 2 (April 1987), p. 76.

"The Earth's Magnetotail" by Edward Hones, Jr., *Scientific American*, Vol. 254, No. 3 (March 1986), p. 40.

"Trace Gases, CO_2, Climate and the Greenhouse Effect" by Gordon J. Aubrecht, II, *Physics Teacher*, Vol. 26, No. 3 (March 1988), p. 145.

"Solar Energy — How Much Do We Receive?" by Uri Ganiel and Oved Kedem, *Physics Teacher*, Vol. 21, No. 9 (December 1983), p. 573.

"Solar Radiation and the Earth's Atmosphere" by William Kuhn and Susan Postawko, *Physics Teacher*, Vol. 26, No. 5 (May 1980), p. 266.

"The Sunspot Cycle: Tip of the Iceberg" by Leif Robinson, *Sky and Telescope*, Vol. 73, No. 6 (June 1987), p. 589.

"The Atmosphere" by Andrew P. Ingersoll, *Scientific American*, Vol. 249, No. 3 (September 1983), p. 162.

"The Sun's Influence on the Earth's Atmosphere and Interplanetary Space" by J. V. Evans, *Science*, Vol. 216, No. 4545 (April 30, 1982), p. 467.

"Snow-free Rings Around Trees" by Daniel A. Mazzarella, *Weatherwise*, Vol. 32, No. 6 (December 1979), p. 247.

"All That's Best of Dark and Bright" by Craig F. Bohren, *Weatherwise*, Vol, 43, No. 3 (June 1990), p. 160.

"The Greenhouse Effect Revisited" by Craig F. Bohren, *Weatherwise*, Vol. 42, No. 1 (February 1989), p. 50.

ANSWERS

MATCHING

1. g
2. d
3. l
4. b
5. p
6. i
7. n
8. f
9. j
10. m
11. a
12. k
13. e
14. c
15. h
16. o

FILL IN THE BLANK

1. reflected
2. black body
3. none
4. window
5. albedo
6. water vapor and carbon dioxide
7. aurora borealis
8. sunspots
9. infrared radiation
10. 30
11. 0.5, 10
12. radiative equilibrium temperature
13. water vapor

MULTIPLE CHOICE

1. b
2. d
3. b
4. d
5. a
6. a
7. d
8. c
9. b
10. a

TRUE - FALSE

1. T
2. F
3. F
4. T
5. T
6. F
7. T
8. F
9. T
10. F
11. F
12. T
13. T
14. F
15. T

PROBLEMS

1. a. 0.25 micrometers
   b. ultraviolet region
   c. 2^4 or 16 times
2. a. answer a'
   b. answer a
   c. answer b
   d. answer a'
   e. The sun's energy warms the surface and the surface temperature rises. As a result, the higher surface temperature causes a greater emission of infrared radiation. As long as incoming solar energy exceeds outgoing surface energy, the surface temperature will rise. When outgoing surface energy equals incoming solar energy, a balance is reached marking the highest temperature for the day.

CHAPTER 4

SEASONAL AND DAILY TEMPERATURES

The first part of Chapter Four discusses seasonal temperature variations in the Northern and Southern Hemispheres. Here we learn that the seasons are caused by the earth being tilted on its axis as it revolves around the sun. Next is a section that describes how seasonal variations in solar energy can influence temperatures on a much smaller scale, such as on the north and south side of a hill. This leads to a discussion of how air temperatures vary on a daily basis. Here we learn that the daily variation in air temperature near the surface is controlled mainly by the input of solar energy and the output of energy from the surface. The next section considers the significance of temperature variations in terms of practical applications to everyday living. At this point we see that temperature information can influence our lives in many ways, from what clothes we take on a trip to providing critical information for energy-use predictions and agricultural planning. The chapter concludes with a section on instruments that measure temperature.

Some important concepts of this chapter:

1. The seasons are caused by the earth being tilted on its axis as it revolves around the sun. This causes seasonal variations in both the length of daylight and the intensity of sunlight that reaches the surface.
2. The Northern Hemisphere experiences winter when the earth is closest to the sun, while the Southern Hemisphere experiences winter when the earth is farthest from the sun.
3. When the Northern Hemisphere experiences winter (Dec., Jan., Feb.), the Southern Hemisphere experiences summer (and vice versa).
4. South-facing sides of hills tend to be warmer and drier than their north-facing counterparts.
5. During the day, the surface of the earth and the air above it will continue to warm as long as incoming energy (mainly sunlight) exceeds outgoing heat energy from the surface.
6. During the night, the earth's surface cools by giving up more infrared radiation than it receives — a process called radiational cooling.
7. The lowest temperatures during the night and early morning hours are usually observed at the earth's surface. The cold air next to the ground produces a radiation inversion.
8. During the summer, humid climates tend to have lower daytime maximum temperatures and higher nighttime minimum temperatures than do drier climates.

9. Even though two cities may have similar average annual temperatures, the range and extreme of their temperatures can vary greatly.
10. The wind-chill factor is a measure of the cooling effect of the wind on the human body. It *does not* relate to inanimate objects such as water pipes, etc.
11. All thermometers must, in some way, be shielded from the sun when measuring air temperature, otherwise their readings will be inaccurate.

SELF TESTS

MATCH THE FOLLOWING

______	1. The astronomical beginning of fall	a. radiation inversion
______	2. Warmer hillsides that are less likely to experience freezing conditions	b. autumnal equinox
______	3. The day with the fewest hours of daylight in the Northern Hemisphere	c. instrument shelter
______	4. A recording thermometer	d. hypothermia
______	5. Lines of equal temperature	e. radiometer
______	6. Used as an index for fuel consumption	f. summer solstice
______	7. The day when, at noon, the sun is at its highest position in the Northern Hemisphere	g. growing degree-days
______	8. Thermometer with a small constriction just above the bulb	h. minimum thermometer
______	9. The rapid lowering of human body temperature can produce this	i. vernal equinox
______	10. Obtains air temperature by measuring emitted infrared energy	j. heating degree-day
______	11. A measured increase in air temperature just above the ground	k. thermograph
______	12. Abbreviation for incoming solar radiation	l. insolation
______	13. Thermometer most likely to contain alcohol	m. maximum thermometer
______	14. The astronomical beginning of spring	n. winter solstice
______	15. These mix the air next to the ground by setting up convection currents	o. orchard heaters
______	16. Used as a guide to planting and for determining the approximate date when crops will be ready for harvesting	p. isotherms
______	17. Protects instruments from the weather elements	q. thermal belts

■ FILL IN THE BLANK

1. The difference between the highest and lowest temperatures for any given day is called the daily __________ of temperature.

2. The __________ is the latitude at which days and nights are always of equal length.

3. The sun will be directly above Honolulu, Hawaii (latitude 21°N) __________ time(s) each year.

4. On a clear, calm night, the ground and air above cool mainly by this process: __________ __________.

5. Solar panels on a solar home built in the Northern Hemisphere should face toward this direction: __________.

6. At the North Pole, the sun rises above the horizon on March 21 and stays above the horizon until this date: __________.

7. The difference between the average temperature of the warmest and coldest month is called the annual __________ of temperature.

8. During January, it is winter in the Northern Hemisphere, and __________ in the Southern Hemisphere.

9. An unseasonably warm spell with clear weather usually near the middle of autumn is often called __________ __________.

10. In the middle latitudes of the Northern Hemisphere, on the summer solstice, the sun rises in the northeast and sets in the __________.

11. List the four necessary conditions for the development of a strong radiation inversion: ______________, ______________, ______________, ______________.

■ MULTIPLE CHOICE

1. In clear weather the air next to the ground is usually __________ than the air above during the night, and __________ than the air above during the day.

   a. colder, warmer
   b. colder, colder
   c. warmer, colder
   d. warmer, warmer

2. In the middle latitudes of the Northern Hemisphere on December 22, the sun may be described as:

   a. rising in the southeast and setting in the northwest
   b. rising in the northeast and setting in the northwest
   c. rising in the northeast and setting in the southwest
   d. rising in the southeast and setting in the southwest

3. The wind-chill factor:

   a. determines how low the air temperature will be on any given day
   b. tells farmers when to protect crops from a freeze
   c. takes into account humidity and air temperature in expressing the current air temperature
   d. indicates the temperature at which water freezes on exposed skin
   e. relates body heat loss with wind to an equivalent temperature with no wind

4. Which below is *not* a primary reason for the seasons in the middle latitudes of the Northern Hemisphere?

   a. the closeness of the earth to the sun
   b. the angle at which sunlight reaches the earth
   c. the length of daylight hours

5. The primary cause of a radiation inversion is:

   a. infrared radiation absorbed by the earth's surface
   b. infrared radiation absorbed by the atmosphere and clouds
   c. infrared radiation emitted by the earth's surface
   d. solar radiation absorbed by the earth's surface
   e. solar radiation reflected by the earth's surface

6. Annually, polar regions lose more heat energy than they receive, yet they are prevented from becoming progressively colder each year mainly by the:

   a. absorption of heat by snow and ice surfaces
   b. circulation of heat by the atmosphere and oceans
   c. conduction of heat through the interior of the earth
   d. storage of heat in the soil beneath the snow cover
   e. release of sensible heat to the atmosphere when the polar ice melts

7. The largest annual range of temperatures is found:

   a. in polar latitudes over land
   b. in polar latitudes over water
   c. at the equator
   d. in middle latitudes near large bodies of water
   e. in the northern Central Plains of the United States

8. If tonight's air temperature is going to drop into the middle 20's (°F) and a fairly stiff wind is predicted, probably the best way to protect an orchard against a hard freeze is to (cost is not a factor):

   a. use wind machines
   b. use helicopters
   c. put orchard heaters to work
   d. sprinkle the trees with water
   e. pray for clouds

9. The earth is tilted at an angle of $23^1/_2$°. If the amount of tilt were decreased to 5°, we would expect to observe in the middle latitudes of the Northern Hemisphere:

   a. warmer summers and colder winters than at present
   b. cooler summers and colder winters than at present
   c. cooler summers and warmer winters than at present
   d. warmer summers and warmer winters than at present
   e. no appreciable change from present conditions

10. Suppose yesterday morning you noticed ice crystals (frost) on the ground, yet the minimum temperature reported in the newspaper was only 35°F. The *most* likely reason for this apparent discrepancy is that:

    a. the temperature reading was taken in an instrument shelter more than 5 ft. above the ground
    b. the thermometer was in error
    c. the newspaper reported the wrong temperature
    d. the thermometer was read before the minimum temperature was reached for the day
    e. the thermometer was read incorrectly

11. An important reason for the large daily temperature range over deserts is:

    a. the light colored sand radiates heat very rapidly at night
    b. dry air is a very poor heat conductor
    c. there is little water vapor in the air to absorb and reradiate infrared radiation
    d. the ozone content of desert air is very low
    e. free convection cells are unable to form over the hot desert ground

12. The most important reason why summers in the Southern Hemisphere are not warmer than summers in the Northern Hemisphere is that:

    a. the earth is farther from the sun during the Southern Hemisphere summer
    b. the Southern Hemisphere is cloudier during its summer
    c. a greater percentage of the Southern Hemisphere is covered with water

13. The adjacent diagram represents the average amount of solar energy received at the earth's surface on what day?

    a. summer solstice — June 22
    b. winter solstice — Dec. 22
    c. vernal equinox — Mar. 21

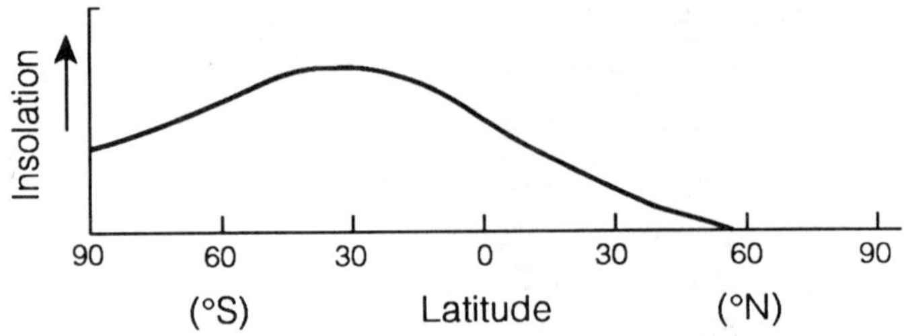

Figure 1

■ TRUE - FALSE

______ 1. Hypothermia is most common in cold, wet weather.

______ 2. The greatest variation in daily temperature normally occurs at the ground.

______ 3. One reason why water warms and cools more slowly than land is because water has a lower heat capacity.

______ 4. An ordinary liquid-in-glass thermometer held in direct sunlight will always indicate a temperature higher than the true air temperature.

______ 5. In most areas, the warmest time of the day about 5 feet above the ground occurs around noon.

______ 6. During the summer, humid regions typically have lower daily temperature ranges and lower maximum temperatures than do drier regions.

______ 7. The earth comes closer to the sun in January than it does in July.

______ 8. One reason why summers are warmer than winters in middle latitudes is that during the summer middle latitudes receive more direct (more intense) sunlight than during the winter.

______ 9. During the afternoon, the greatest temperature difference between the surface air and the air several feet above occurs on clear, calm afternoons.

______ 10. At Barrow, Alaska, there are fewer hours of sunlight during the middle of July than there are at the end of June.

______ 11. If two cities have the same mean annual temperature, then their temperatures throughout the year must be quite similar.

______ 12. When the air temperature is 35°F and the wind is blowing at 30 miles per hour, the wind chill factor will be below freezing and exposed water pipes will likely freeze.

_____ 13. Incoming sunlight in middle latitudes is less in winter than in summer partly because the sun's rays slant more and spread their energy over a larger area.

_____ 14. Radiation temperature inversions are best developed in the early morning just before sunrise.

_____ 15. In the middle latitudes of the Northern Hemisphere, between Christmas and New Years, the length of daylight increases each day.

_____ 16. If you travel from Dallas, Texas to St. Paul, Minnesota during July, you are more likely to experience greater temperature variations than if you made the same trip in January.

_____ 17. In a hilly region, the best place to plant crops that are sensitive to low temperatures is on the valley floor.

■ PROBLEMS

1. The adjacent diagram is a section of a minimum thermometer.

   a. the current air temperature is _________°F

   b. the minimum temperature is _________°F

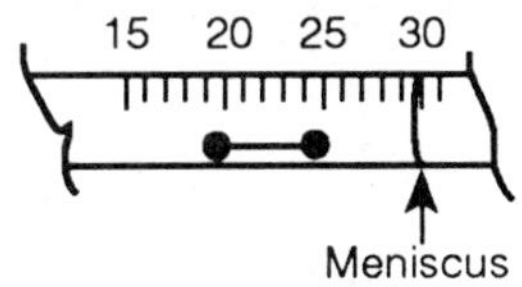

Figure 2

2. How many heating degree-days would there be in a city when the maximum temperature is 40°F and the minimum temperature is 20°F? (Assume a base temperature of 65°F)

3. How many cooling degree-days would there be in a city when the maximum temperature is 95°F and the minimum temperature is 75°F? (Assume a base temperature of 65°F.)

4. Suppose corn is planted in Maryland on March 1. If this variety of corn needs 2400 growing degree-days before it can be picked, and if the mean daily temperature for March through August is 65°F, in about how many days would the corn be ready to pick? On about what date would this be? (Assume a base temperature of 50°F.)

Pick corn in about _________ days.

The date of picking will be about _________.

5. Since New York City is at about $40^1/_2$°N latitude, calculate how high the sun is above the southern horizon on the winter solstice and on the summer solstice. (Information on this is given on p. 91 of your text.)

Answer for Dec. 22 _________

Answer for June 22 _________

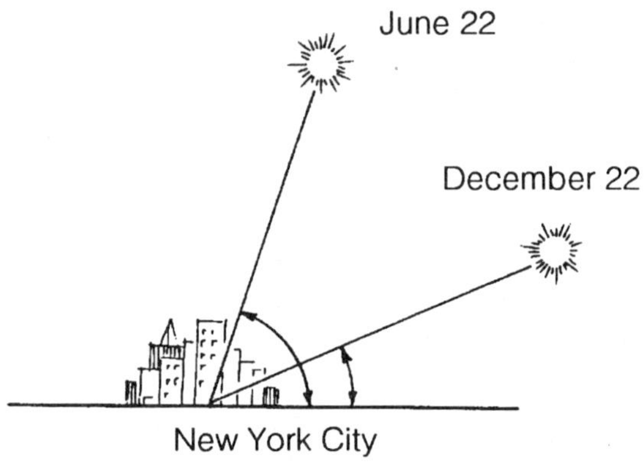

Figure 3

6. In Fig. 4.21, p. 102 of your text, determine the approximate wind chill under the following conditions:

a. air temperature -10°F, wind speed 20 mi/hr Answer _________

b. air temperature -35°C, wind speed 10 knots Answer _________

7. The adjacent diagram (Fig. 4) represents sunrise on three different days. One day is the autumnal equinox (Sept. 23), one is the summer solstice (June 22), and the third is the winter solstice (Dec. 22).

   a. In the space above each sun, place the proper date.

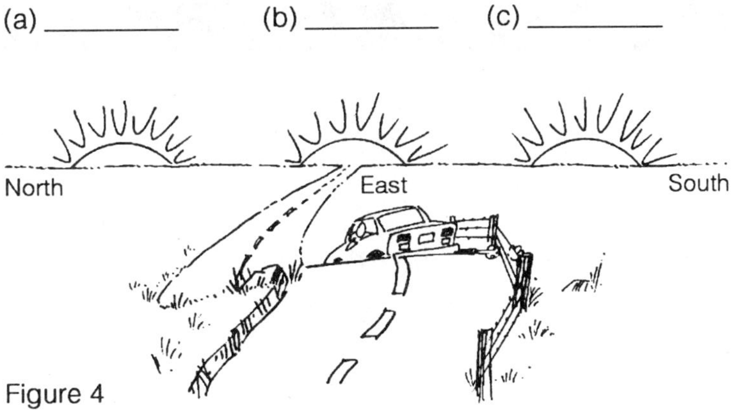

Figure 4

   b. Look back through old newspapers (and court records if available) to see if there is a propensity for accidents (both auto and pedestrian) near sunrise around the autumnal equinox in your area. If there is, explain why with the aid of Figure 4.

■ ADDITIONAL READINGS

"Weather Records: The Case of the Bennett, Colorado, Maximum Temperature" by Thomas W. Bettge, *Weatherwise*, Vol. 38, No. 2 (April 1985), p. 95.

"Windchill: The 'Burr' Index" by Dennis M. Driscoll, *Weatherwise*, Vol. 40, No. 6 (December 1987), p. 321.

"Degree Days: Heating and Cooling by the Numbers" by J. Murray Mitchell, *Weatherwise*, Vol. 40, No. 6 (December 1987), p. 334.

"Temperature Inversions Have Cold Bottoms" by Craig F. Bohren and Gail M. Brown, *Weatherwise*, Vol. 34, No. 6 (December 1981), p. 273.

"The Great Freeze of '83 and the Effects" by George A. Winterling, *Weatherwise*, Vol. 37, No. 6 (December 1984), p. 304.

"Astronomical vs. Meteorological Winter" by Thomas Schlatter, *Weatherwise*, Vol. 38, No. 1 (February 1985), p. 42.

"Value of Weather Information: A Descriptive Study of the Fruit-Frost Problem" by Thomas R. Stewart, Richard W. Katz and Allen H. Murphy, *Bulletin of the American Meteorological Society*, Vol. 65, No. 2 (February 1984), p. 126.

ANSWERS

MATCHING

1. b
2. q
3. n
4. k
5. p
6. j
7. f
8. m
9. d
10. e
11. a
12. l
13. h
14. i
15. o
16. g
17. c

FILL IN THE BLANK

1. range
2. equator
3. two
4. radiational cooling
5. south
6. Sept. 23 or autumnal equinox
7. range
8. summer
9. Indian summer
10. northwest
11. a. clear
    b. calm
    c. dry (little water vapor)
    d. long winter night

MULTIPLE CHOICE

1. a
2. d
3. e
4. a
5. c
6. b
7. a
8. d
9. c
10. a
11. c
12. c
13. b

TRUE - FALSE

1. T
2. T
3. F
4. T
5. F
6. T
7. T
8. T
9. T
10. T
11. F
12. F
13. T
14. T
15. T
16. F
17. F

PROBLEMS

1. a. about 30°
   b. about 25°
2. 35 heating-degree days
3. 20 cooling-degree days
4. about 160 days; The date of picking would be around August 7
5. Dec. 22: 26°
   June 22: 73°
6. a. about -55°F
   b. about -51°C
7. a. (a) summer solstice
   (b) autumnal equinox
   (c) winter solstice
   b. On the autumnal equinox, looking east at sunrise would find the sun directly in the eyes of a driver.

CHAPTER 5

LIGHT, COLOR AND ATMOSPHERIC OPTICS

Chapter Five focuses on sunlight in our atmosphere and the visual effects it produces. The chapter begins by describing how scattered sunlight can produce an array of atmospheric visuals from white clouds to blue skies to crepuscular rays. The next several sections discuss how the bending of light produces such phenomena as mirages, halos, and sundogs. After this we see that rainbows are the result of light being bent, reflected, and dispersed. The later part of the chapter describes how coronas, glories, and cloud iridescence form.

Some important concepts of this chapter:

1. Light that is scattered is sent in all directions — forward, sideways, and backwards.

2. An object that appears white, yet is not hot enough to generate its own light, has all visible wavelengths reflected or scattered from its surface. An object that technically has little or no visible light returning from its surface appears black.

3. Blue skies are the result of air molecules scattering the shorter waves of visible light (blue) more than the longer visible waves (red).

4. The bending of light as it moves through regions (objects) of different density is called refraction.

5. The dispersion of light separates white light into its different component colors.

6. For a rainbow to form, rain must be falling in one part of the sky and the sun must be shining in another.

7. The bending of light as it passes around objects is called diffraction.

SELF TESTS

MATCH THE FOLLOWING

______ 1.	You can see this only when the sun is to your back and it is raining in front of you	a.	halo
______ 2.	Phenomenon that makes objects appear higher or lower than they actually are	b.	glory
______ 3.	The time after sunset or before sunrise when the sky is illuminated	c.	green flash
______ 4.	Another name for diffuse light	d.	mirage
______ 5.	A ring of light encircling and extending outward from the sun or moon	e.	corona
______ 6.	Beams of light shining downward through breaks in clouds	f.	Heiligenschein
______ 7.	A white ring surrounding the shadow of an observer's head on a dew-covered lawn	g.	rainbow
______ 8.	Colored rings around the shadow of an aircraft	h.	scattered light
______ 9.	A vertical streak of light extending above (or below) the sun	i.	sun pillar
______ 10.	A ring of light that appears to rest on the moon or sun	j.	scintillation
______ 11.	This can be seen near the upper rim of the sun at sunrise or sunset	k.	crepuscular rays
______ 12.	The apparent twinkling of a star	l.	twilight

ADDITIONAL MATCHING

Match the following with the process that is mainly responsible for its formation. (Note: some answers will be used more than once.)

______ 1. crepuscular rays	a.	scattering
______ 2. inferior mirage	b.	reflection
______ 3. glory	c.	refraction
______ 4. halo	d.	diffraction
______ 5. blue skies		
______ 6. tangent arc		
______ 7. cloud iridescence		
______ 8. sundog		
______ 9. sun pillar		
______ 10. corona		
______ 11. white clouds		
______ 12. star scintillation		
______ 13. brocken bow		
______ 14. hazy skies		
______ 15. blue moons		
______ 16. wet-looking pavement on a clear, dry, hot day		

FILL IN THE BLANK

1. If there was no atmosphere surrounding the earth, the *sky* during the day would appear what color? __________

2. When sunlight strikes an object and the light is sent in all directions, the light is said to be __________.

3. The bending of light through an object is called __________.

4. The breaking up of white light by selective refraction is called __________.

5. The type of mirage that makes objects appear higher than they actually are is called a __________ __________.

6. The bending of light as it passes around an object is called __________.

7. For halos, tangent arcs, and sun pillars to form, __________ __________ must be present in the atmosphere.

MULTIPLE CHOICE

1. If you were standing on the moon, the color of the *sky* would be:
   a. white
   b. red
   c. blue
   d. purple
   e. black

2. Which of the following processes must occur in a raindrop to produce a rainbow?
   a. refraction, reflection, and scattering of sunlight
   b. reflection, scattering, and dispersion of sunlight
   c. refraction, reflection, and dispersion of sunlight
   d. transmission, scattering, and dispersion of sunlight
   e. refraction, transmission and scattering of sunlight

3. As light passes through ice crystals, __________ light is bent the *least* and is, therefore observed on the __________ of halos and sundogs.
   a. red, outside
   b. red, inside
   c. blue, inside
   d. blue, outside

4. On a foggy night it is usually difficult to see the road when the high beam lights are on because of the _________ of light.

   a. scattering
   b. absorption
   c. transmission
   d. refraction
   e. diffraction

5. If you looked in the general direction of the sun you would not be able to see a:

   a. corona
   b. halo
   c. sundog
   d. rainbow
   e. sun pillar

6. If the setting sun appears red, you may conclude that:

   a. the sun's surface temperature has changed
   b. the next day will be extremely hot
   c. only the longest waves of visible light are striking your eye
   d. the moon will not shine
   e. there is something in your eye

7. Clouds in the tropics tend to move from east to west. Consequently, which rhyme best describes a rainbow seen in the tropics?

   a. Rainbow at the break of day/ means that rain is on the way
   b. Rainbow in morning/ sailors should take warning
   c. Rainbow with a setting sun/ means that sailors can have some fun
   d. Rainbow at the break of dawn/ means, of course, the rain is gone

8. When distant mountains appear blue, it is mainly due to the _________ of light.

   a. scattering
   b. dispersion
   c. refraction
   d. diffraction
   e. scintillation

TRUE - FALSE

_______ 1. Rainbows form in much the same way as a mirage.

_______ 2. At sunrise in the middle latitudes, a sundog would appear in the west.

_______ 3. The Fata Morgana is actually a type of mirage.

_______ 4. When a star appears near the horizon, its actual position is slightly higher.

_______ 5. The best time of day to see a rainbow is around noon.

_______ 6. If the earth had no atmosphere, the stars would be visible all the time.

_______ 7. Small suspended salt particles, volcanic ash, and small suspended dust particles are all capable of producing red sunrises and sunsets.

_______ 8. Secondary rainbows occur when two internal reflections of light occur in raindrops.

_______ 9. The Heiligenschein is most easily seen at night.

PROBLEMS

1. In the space provided below, place the name of the optical phenomenon that appears in the adjacent figures.

a. ____________________

b. ____________________

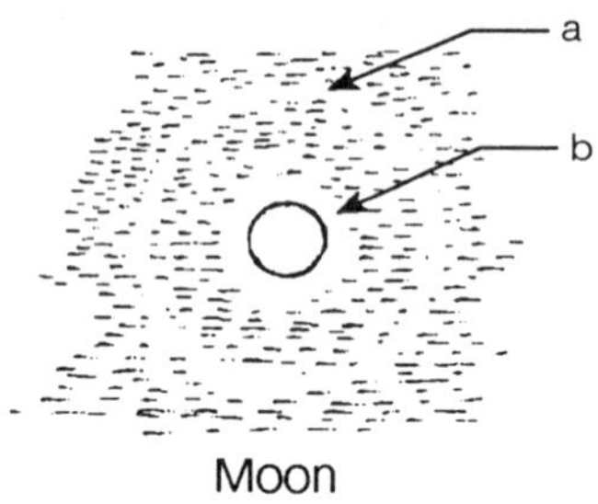

Figure 1

c. ____________________

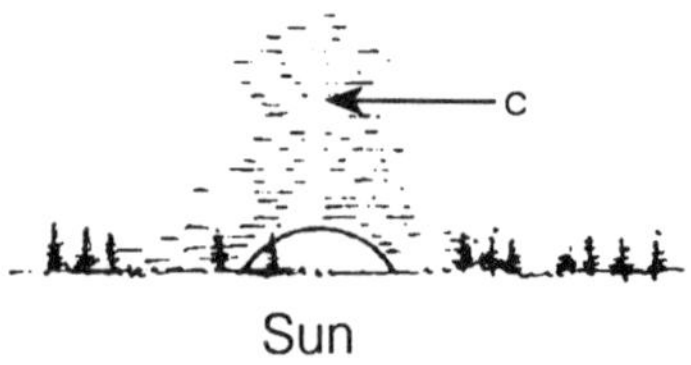

Figure 2

d. ______________________

e. ______________________

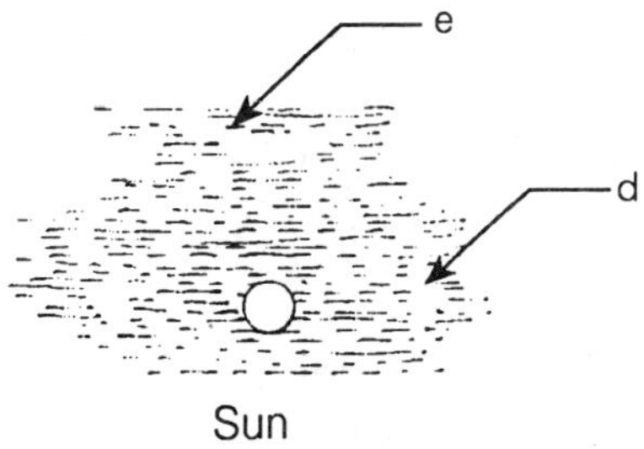

Figure 3

f. ______________________

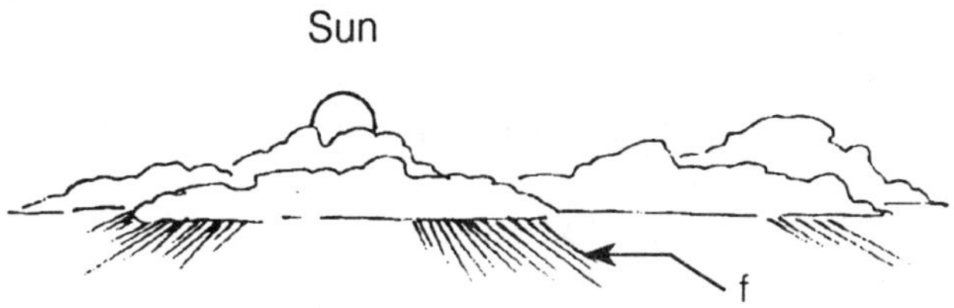

Figure 4

g. ______________________

Figure 5

h. ______________________

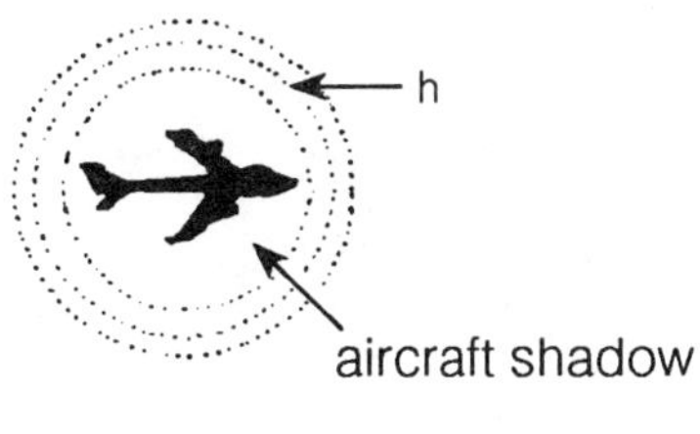

Figure 6

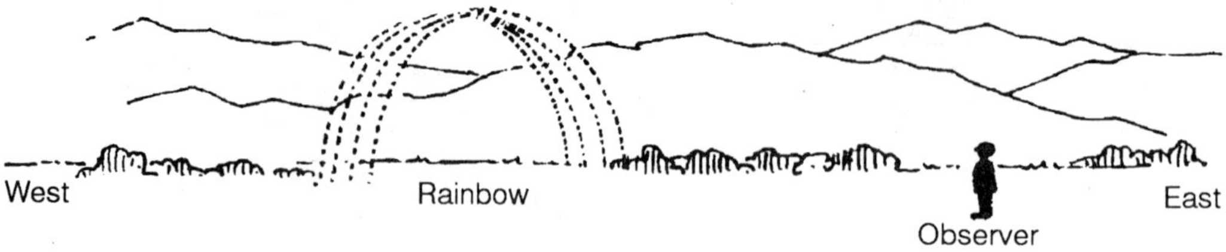

Figure 7

2. In Figure 7:

   a. Place the sun in its proper position for the rainbow to appear as shown.

   b. Place clouds and falling rain in their proper position.

   c. If the person observing the rainbow is located in the Central Plains of the United States, based on the information in the illustration, would the person expect clearing weather or showers? Explain.

ADDITIONAL READINGS

"Colors of the Sky" by Craig F. Bohren and Alistair Fraser, *Physics Teacher*, Vol. 23, No. 5 (May 1985), p. 267.

"Chasing Rainbows" by Alistair B. Fraser, *Weatherwise*, Vol. 36, No. 6 (December 1983), p. 280.

"The Boulder, Colorado, Concentric Halo Display of 21 July 1986" by Paul J. Neiman, *Bulletin of the American Meteorological Society*, Vol. 70, No. 3 (March 1989), p. 258.

Rainbows, Halos, and Glories by Robert Greenler; Cambridge University Press, New York, Cambridge, 1980.

Clouds in a Glass of Beer by Craig F. Bohren; John Wiley and Sons, Inc., New York, 1987. See especially pp. 86-185.

"Light Pillar Climatology" by Kenneth Sassen, *Weatherwise*, Vol. 33, No. 6 (December 1980), p. 251.

"Highway Mirages" by Craig F. Bohren, *Weatherwise*, Vol. 42, No. 4 (August 1989), p. 224.

"Sundogs" by T. D. Nicholson, *Natural History*, (February 1991), p. 70.

ANSWERS

MATCHING

1. g	4. h	7. f	10. e
2. d	5. a	8. b	11. c
3. l	6. k	9. i	12. j

ADDITIONAL MATCHING

1. a	5. a	9. b	13. d
2. c	6. c	10. d	14. a
3. d	7. d	11. a	15. a
4. c	8. c	12. c	16. c

FILL IN THE BLANK

1. black
2. scattered
3. refraction
4. dispersion
5. superior mirage
6. diffraction
7. ice crystals

MULTIPLE CHOICE

1. e	3. b	5. d	7. d
2. c	4. a	6. c	8. a

TRUE - FALSE

1. F	4. F	6. T	8. T
2. F	5. F	7. T	9. F
3. T			

PROBLEMS

1. a. halo
   b. corona
   c. sun pillar
   d. sundog
   e. tangent arc
   f. crepuscular rays
   g. inferior mirage
   h. glory

2. a. Sun's position would be to the east of the observer.
   b. The clouds and falling rain would be located directly above the rainbow.
   c. Observer could expect showers because at this location clouds tend to move from west to east.

CHAPTER 6

HUMIDITY

Chapter Six examines the many ways of describing and measuring humidity. The chapter begins by looking at absolute humidity, specific humidity, and mixing ratio. It then looks at humidity in terms of vapor pressure and saturation vapor pressure. Next, we learn that although relative humidity is the most common way to describe atmospheric moisture, it is also the most misunderstood. After a discussion on heat indexes, we learn that a good indicator of the air's actual water vapor content is the dew-point temperature. The chapter concludes by describing the different ways of measuring humidity.

Some important concepts of this chapter:

1. Absolute humidity describes the mass of water vapor in a fixed volume of air, or the water vapor density.

2. The actual vapor pressure of air expresses the amount of water vapor in terms of the amount of pressure that the water vapor molecules exert.

3. Relative humidity, expressed as a percent, does not tell us how much water vapor is actually in the air; rather, it tells us how close the air is to being saturated.

4. Without changing the air's water vapor content, as air cools the relative humidity increases, and as air warms the relative humidity decreases.

5. The dew-point temperature is a good indicator of the air's water vapor content. High dew points indicate high water vapor content and vice versa.

6. When the air temperature and dew point are close together, the relative humidity is high; when they are far apart, the relative humidity is low.

7. Summertime is generally more humid in the eastern half of the United States because of the air flow off the warm Gulf of Mexico.

8. High relative humidities in hot weather can make us feel it is hotter than it actually is by retarding the evaporation of perspiration.

SELF TESTS

MATCH THE FOLLOWING

______ 1. The density of water vapor in a given volume of air

______ 2. Measures relative humidity with human hairs

______ 3. Combines air temperature with relative humidity to determine an apparent temperature

______ 4. Uses wet-bulb and dry-bulb temperatures to obtain relative humidity

______ 5. The maximum pressure that water vapor would exert if the air were saturated

______ 6. Measures humidity by measuring the amount of infrared energy absorbed by water vapor

______ 7. Used to cool the air in hot, dry climates

______ 8. The mass of water vapor in a given mass of air

a. infrared hygrometer

b. Heat Index

c. hair hygrometer

d. saturation vapor pressure

e. specific humidity

f. absolute humidity

g. evaporative cooler

h. sling psychrometer

FILL IN THE BLANK

1. __________ __________ can be described as the percent of water vapor in the air compared to that required for saturation.

2. The lowest temperature that can be attained by evaporating water into the air is the __________ __________.

3. The temperature to which air must be cooled for saturation to occur is called the __________ __________.

4. On most days the relative humidity reaches its highest value when the air temperature reaches its __________ value.

■ MULTIPLE CHOICE

1. When the air temperature increases, the saturation vapor pressure:

   a. increases
   b. decreases
   c. does not change

2. The instrument that determines the amount of water vapor in the air by measuring the air's actual vapor pressure is the:

   a. sling psychrometer
   b. electrical hygrometer
   c. dew-cell
   d. hair hygrometer

3. Which is the *best* indicator of the actual amount of water vapor in the air?

   a. air temperature
   b. dew-point temperature
   c. relative humidity
   d. wet-bulb temperature
   e. saturation vapor pressure

4. If the air temperature remains constant, evaporating water into the air will __________ the dew-point temperature and __________ the relative humidity.

   a. decrease, decrease
   b. decrease, increase
   c. increase, decrease
   d. increase, increase

5. Which best explains why in North America the Gulf Coast states are more humid in summer than the coastal areas of southern California?

   a. the surface water temperature of the Gulf of Mexico is very warm
   b. the air temperatures are higher over the Gulf States
   c. the air over the Pacific ocean has a much lower relative humidity

6. Polar air is considered "dry" because the dew-point temperatures are often quite low. However, the relative humidity of this cold, polar air is usually high because:

   a. low dew points indicate that the relative humidity must be high
   b. low air temperatures indicate that the relative humidity must be high
   c. the air temperature and the dew point are fairly close together

TRUE - FALSE

______ 1. The relative humidity is a measure of the air's actual water vapor content.

______ 2. The temperature at which water boils is dependent mainly upon air temperature.

______ 3. On a hot, humid day, a good measure of how cool the human skin can become is the wet-bulb temperature.

______ 4. Relative humidity is always given as a percent.

______ 5. The ratio of the mass of water vapor in a given volume of air to the mass of remaining dry air describes the mixing ratio.

______ 6. If you turn on your oven and open its door, the increase in air temperature should raise the relative humidity inside your home.

______ 7. When the air is saturated, an increase in air temperature will cause condensation to occur.

______ 8. Suppose the air temperature inside your home is 78°F and you lower it to 68°F. As long as the moisture content of the air inside does not change, the relative humidity should increase.

PROBLEMS

1. List the two most important factors that cause the relative humidity of air to change.

   a.

   b.

TABLE 1

Saturation Vapor Pressure

Temperature °F	Vapor Pressure, mb	Temperature °F	Vapor Pressure, mb	Temperature °F	Vapor Pressure, mb
-20°	0.6	30°	5.6	80°	35.0
-15°	0.7	35°	6.9	85°	41.0
-10°	0.9	40°	8.4	90°	48.1
-5°	1.2	45°	10.2	95°	56.2
0°	1.5	50°	12.3	100°	65.6
5°	1.9	55°	14.8	105°	76.2
10°	2.4	60°	17.7	110°	87.8
15°	3.0	65°	21.0	115°	101.4
20°	3.7	70°	25.0	120°	116.8
25°	4.6	75°	29.6	125°	134.2

2. Listed below are three cities, each with their air temperature and dew-point temperature for 3 PM on a given summer day.

	Air Temperature	Dew-point Temperature
City A	105°F	75°F
City B	75°F	60°F
City C	85°F	65°F

a. Which city has the highest relative humidity?

b. Which city has the lowest relative humidity?

c. Which city has the greatest amount of water vapor in the air?

d. Which city has the least amount of water vapor in the air?

e. Use Table 1 (Saturation Vapor Pressure) and the formula,

$$\text{relative humidity} = \frac{e}{e_s} \times 100\%,$$

to obtain the relative humidity for each of the three cities. (Additional information on obtaining relative humidity using this formula is given in your textbook on p. 147.)

Relative humidity for City A __________%

Relative humidity for City B __________%

Relative humidity for City C __________%

f. Use the air temperatures for each city, and the relative humidities you calculated in (e), to obtain the Heat Index (HI)/ apparent temperature. To obtain the apparent temperature use the graph in Fig. 6.10, p. 144 of your text. For each apparent temperature, observe the heat syndrome that goes along with it in Table 6.2, p. 144 of your text.

	Apparent Temperature (°F)	Category Number
City A		
City B		
City C		

g. Now, check your apparent temperatures by using the Heat Index, Table 1.1 on p. 545 in your text.

Apparent temperature City A __________ °F

Apparent temperature City B __________ °F

Apparent temperature City C __________ °F

3. A 90/90 problem.

On extremely humid days in the eastern half of North America, a common complaint is that the air temperature is 90°F and the relative humidity is 90 percent. Is this really possible?

a. To find out, first calculate what the actual vapor pressure (e) must be for a relative humidity of 90% with an air temperature of 90°F. Use the formula for relative humidity given in problem 2e, and the vapor pressure given in Table 1. (Remember, additional information on the use of the formula is given on p. 147 of your textbook.)

Actual Vapor Pressure, e, = __________ mb

b. Now, to determine what the dew-point temperature must be with an air temperature of 90°F and relative humidity of 90 percent, simply find in Table 1 (vapor pressure table) the dew-point temperature that corresponds to the actual vapor pressure you calculated in (a).

Dew-point temperature with 90°F air temperature and 90 percent relative humidity = __________ °F.

c. In your textbook on p. 138, Fig. 6.6, notice that even in the humid Gulf States, the average vapor pressures for July are *much* lower than that which you calculated for a day with an air temperature of 90°F and a relative humidity of 90 percent. Consequently, although it is remotely possible, it is extremely improbable that you will ever experience a day when the air temperature is 90° F and the relative humidity is 90 percent.

4. Use Figure 6.6, p. 138 in your textbook to determine the approximate actual vapor pressures for the following cities during July:

CITY	Average Vapor Pressure July, mb	Average Dew Point °F	Average Dew Point °C
Philadelphia, PA			
Denver, CO			
New Orleans, LA			
Fargo, ND			

To determine the average July dew-point temperature for each city, find in Table 1 (vapor pressure table) the dew point temperature that corresponds to the average vapor pressure for each city.

In the last column, convert the dew-point readings to °C.

Figure 2

5. a. In Figure 2, the little house is situated on a cold, snowy prairie. The outside air temperature and dew point for this day are shown. What is the relative humidity of the outside air?

Relative humidity of outside air ________%

b. Suppose the window in the little house is open just a crack and the cold outside air flows inside. If no water vapor is either added to or removed from the outside air as it blows into the house, what would be the dew-point temperature of the air inside? ________°F

Further suppose that the radiator inside the house warms the air to a comfortable 70°F. What would be the relative humidity of the air inside this little house? (Hint: use the formula for calculating relative humidity given in problem 2e and the vapor pressures given in Table 1.)

Relative humidity of inside air ________%

6. a. With a sling psychrometer you measure an air temperature of 95°F and a wet-bulb temperature of 77°F. With the aid of Table D.3, p. 529 in your textbook, determine the dew-point temperature. ________°F

b. Using Table D.4, p. 530 in your textbook, determine the relative humidity of this air. ________%

c. Now, use the formula given in 2e and vapor pressures in Table 1 to calculate the relative humidity of the air. (Your answer should be very close to what you determined in problem b.)

Relative humidity = ________%

■ ADDITIONAL READINGS

"How Humid is Humid" by David E . Siskind, *Weatherwise*, Vol. 44, No. 3 (June 1991), p. 24.

"The Summer Simmer Index" by John W . Pepi, *Weatherwise*, Vol. 40, No. 3 (June 1987), p. 143.

"Heat Stress: A Comparison of Indices" by Robert Quayle and Fred Doehring, *Weatherwise*, Vol. 34, No. 3 (June 1981), p. 120.

"Genies in Jars, Clouds in Bottles, and a Bucket with a Hole in it" by Craig F. Bohren and Gail M. Brown, *Weatherwise*, Vol. 35, No. 2 (April 1982), p. 86.

"Water Vapor Mysticism" by Craig Bohren, *Weatherwise*, Vol 43, No. 2 (April 1990), p. 97.

"Boil and Bubble, Toil and Trouble" by Craig F. Bohren, *Weatherwise,* Vol. 42, No . 2 (April 1989), p. 104.

"Sugar and Spice: The Dirty Wet-Bulb Temperature" by Craig F. Bohren, *Weatherwise*, Vol. 39, No. 1 (February 1986), p. 46.

"Building a Dew Point Hygrometer" by Hampton W. Shirer, *Weatherwise*, Vol. 39, No. 2 (June 1986), p. 160.

ANSWERS

MATCHING

1. f
2. c
3. b
4. h
5. d
6. a
7. g
8. e

FILL IN THE BLANK

1. relative humidity
2. wet-bulb temperature
3. dew-point temperature or dew point
4. lowest or minimum

MULTIPLE CHOICE

1. a
2. c
3. b
4. d
5. a
6. c

TRUE - FALSE

1. F
2. F
3. T
4. T
5. T
6. F
7. F
8. T

PROBLEMS

1. a. changes in air temperature
   b. changes in water vapor content
2. a. city B
   b. city A
   c. city A
   d. city B
   e. relative humidity for city A, 39%
   relative humidity for city B, 60%
   relative humidity for city C, 51%
   f. city A: apparent temperature approximately 121°F, category II
   city B: apparent temperature less than 80°F
   city C: apparent temperature approximately 88°F, category IV
   g. city A, about 122°F
   city B, 76°F
   city C, 88°F
3. a. e = 43.3 mb
   b. dew-point temperature is 86.6°F or about 87°F
4. Philadelphia, PA: Avg. vapor pressure July, 21 mb, dew point 65°F (18.3°C).
   Denver, CO: Avg. vapor pressure July, 11 mb, dew point 47°F (8.3°C).
   New Orleans, LA: Avg. vapor pressure July, 27 mb, dew point 72°F (22.2°C).
   Fargo, ND: Avg. vapor pressure July, 15 mb, dew point 55°F (12.8°C).
5. a. RH = 100%
   b. inside dew-point temperature 15°F, inside RH = 12%
6. a. dew-point temperature 70°F
   b. RH = 43%
   c. RH = 44%

CHAPTER

7

CONDENSATION: DEW, FOG, AND CLOUDS

Chapter Seven describes the various forms of condensation. The chapter begins with a discussion on the formation of dew, frost, and fog. This is followed by a section on cloud classification. Here we learn that clouds are classified according to their height above the ground and their physical appearance. The next section helps with the identification of clouds by providing you with many visual clues. The latter part of the chapter illustrates some of the current methods used in observing clouds.

Some important concepts of this chapter:

1. Dew, frost, and frozen dew form when objects on the surface cool below the air's dew-point temperature.

2. Condensation nuclei are important in the atmosphere because they serve as surfaces on which water vapor condenses.

3. Fog can form as the air cools, or as water evaporates and mixes with drier air.

4. Clouds are usually divided into four main groups: high, middle, low, and clouds with vertical development.

5. Satellites not only photograph clouds, they provide scientists with a great deal of physical information about the earth and its atmosphere.

6. A color cloud chart appears at the end of your textbook. Remove it and use it while observing the sky.

SELF TESTS

MATCH THE FOLLOWING

______ 1.	Another name for visible white frost	a.	upslope fog
______ 2.	These particles serve as surfaces on which water vapor may condense	b.	winter chilling
______ 3.	Fog that forms as moist air flows upward along an elevated surface	c.	condensation nuclei
______ 4.	Arctic sea smoke and "steam devils" are a form of this type of fog	d.	ceilometer
______ 5.	A tiny liquid drop of dew that freezes when the air temperature drops below freezing	e.	frontal fog
______ 6.	A positive benefit of fog for fruit and nut trees	f.	advection fog
______ 7.	Fog that most often forms as warm rain falls into a cold layer of surface air	g.	hoarfrost
______ 8.	When fog "burns off" it does this	h.	dew
______ 9.	The most common type of fog along the Pacific coast of North America	i.	frozen dew
______ 10.	Beads of water that have condensed onto objects near the ground	j.	radiation fog
______ 11.	Fog that most commonly forms on clear nights, with light or calm winds	k.	evaporates
______ 12.	Instrument that measures the height of a cloud's base above the ground	l.	steam fog

■ ADDITIONAL MATCHING (CLOUDS)

(Some answers may be used more than once.)

______ 1. A "mackerel sky" describes this cloud

______ 2. A low, lumpy cloud layer that appears in rows, patches, or rounded masses

______ 3. A towering cloud that has not fully developed into a thunderstorm

______ 4. Hail is usually associated with this cloud type

______ 5. The sun or moon are dimly visible or appear watery through this gray, sheetlike cloud

______ 6. A halo around the sun or moon often identifies the presence of this cloud

______ 7. Wispy, high clouds

______ 8. Light or moderate but steady precipitation that covers a broad area is most often associated with this cloud

______ 9. This cloud's elements (puffs) should be about the size of your thumbnail when your hand is extended to arm's length

______ 10. Lightning and thunder are associated with this cloud

______ 11. A cloud of vertical development that resembles a small piece of floating cotton

______ 12. A middle cloud that occasionally forms in parallel waves or bands

______ 13. The cloud with the smallest elements or puffs as viewed from the surface

______ 14. A low, uniform, grayish cloud, whose precipitation is most commonly drizzle

______ 15. This cloud's elements (puffs) should be about the size of your fist when your hand is extended to arm's length

______ 16. When fog lifts above the surface it forms this gray, sheetlike cloud

______ 17. Cloud with the greatest vertical growth

a. cirrus

b. cirrostratus

c. cirrocumulus

d. altostratus

e. altocumulus

f. nimbostratus

g. stratus

h. stratocumulus

i. cumulus

j. cumulus congestus

k. cumulonimbus

FILL IN THE BLANK

1. Cirrus clouds are composed primarily of __________ __________.

2. Fog that forms as relatively warm air moves over a colder surface is called __________ fog.

3. The cooling of the ground to produce dew and frost is mainly the result of __________ cooling.

4. Clouds that appear as baglike sacks hanging from beneath a cloud are called __________.

5. A sky composed of between one-tenth and five-tenths clouds is technically described as __________.

6. A cloudlike stream seen forming behind a jet aircraft is called a(n) __________.

7. Another name for "water-seeking" condensation nuclei is __________.

8. Clouds with a lens shape that often form over and downwind of mountains are called __________.

9. Another name for a "luminous night cloud" is __________ __________.

10. A cloud that sometimes resembles a silken scarf capping the top of a developing cumulus cloud is the __________.

11. __________ clouds form in the stratosphere and are also called mother-of-pearl clouds.

MULTIPLE CHOICE

1. Ragged-looking clouds that drift rapidly with the wind, often beneath a nimbostratus cloud, are called:

   a. nacreous
   b. pileus
   c. castillanus
   d. scud
   e. mammatus

2. Condensation nuclei are important in the atmosphere because:

   a. they filter out sunlight
   b. they are high energy particles that cool the air
   c. they provide most of the minerals found in water
   d. they help to dissipate smog
   e. without them, condensation would not occur naturally in the atmosphere

3. The fog most likely to form on a clear, calm autumn morning above a cold lake is:

   a. radiation fog
   b. evaporation (mixing) fog
   c. upslope fog
   d. advection fog

4. A cloud that resembles "little castles in the sky" is called:

   a. castellanus
   b. scud
   c. mammatus
   d. banner
   e. lenticular

5. A reasonably successful method of dispersing *cold fog* is to:

   a. warm the air so that the fog evaporates
   b. mix the surface air with warmer air above
   c. seed the fog with dry ice
   d. use the technique called turboclair

6. Of the different types of fog listed below, which one does *not* necessarily form in air that is cooling?

   a. advection fog
   b. radiation fog
   c. upslope fog
   d. evaporation (mixing) fog

7. If you are standing outside and notice that the sky is covered with a high, white layered cloud, and you look at the ground and observe your shadow, you may conclude that the cloud overhead is:

   a. altostratus
   b. cirrostratus
   c. stratus
   d. nimbostratus
   e. stratocumulus

8. The highest clouds in our atmosphere are called:

   a. cirrus
   b. altocumulus
   c. noctilucent
   d. cumulonimbus
   e. cumulus congestus

9. If a cloud appears white in a visible satellite picture and gray in an infrared picture, then the cloud would most likely be a:

   a. low cloud
   b. high cloud
   c. middle cloud

10. The wintertime fog that often occurs in the central valley of California is mainly:

   a. advection fog
   b. radiation fog
   c. evaporation (mixing) fog
   d. upslope fog

11. The satellites that are positioned at the highest level above the earth's surface are:

   a. geostationary satellites
   b. polar orbiting satellites
   c. LandSat satellites

12. Which of the clouds listed below is *least* likely to produce precipitation that reaches the ground?

   a. nimbostratus
   b. cumulus congestus
   c. cumulonimbus
   d. stratus
   e. cirrocumulus

13. A cloud that forms in descending air is:

   a. pileus
   b. mammatus
   c. castellanus
   d. stratocumulus
   e. cumulus fractus

14. Which of the statements below is correct about polar orbiting satellites?
    a. they remain above a fixed location over the poles
    b. they remain fixed over a position of constant longitude
    c. they pass directly over the same place on each successive orbit
    d. they follow lines of latitude as they orbit the earth
    e. on each successive orbit, they monitor an area to the west of the previous orbit

15. The relative humidity could well exceed 100 percent without producing fog when:
    a. the dew point is higher than the air temperature
    b. the water vapor is composed of heavy water
    c. there are no condensation nuclei present
    d. there are too many ions in the air
    e. the air temperature is well below freezing

■ TRUE - FALSE

______ 1. Fog can be composed of ice crystals.

______ 2. In middle latitudes, high clouds are typically observed above altitudes of 20,000 ft (6000 m).

______ 3. Advection fog is more likely to form at the headlands, rather than at the beaches of an irregular coastline because at the headlands surface winds tend to converge and rise.

______ 4. Cirrus clouds appear white in a visible satellite picture and gray in an infrared satellite picture.

______ 5. Frost forms when water vapor changes directly into ice without first becoming a liquid.

______ 6. Dew is most likely to form on clear, windy nights.

______ 7. When clouds are viewed near the horizon, the individual cloud elements often appear farther apart than they actually are.

______ 8. Infrared satellite pictures are computer enhanced to increase the contrast between specific features in the picture.

______ 9. Valleys are more susceptible to radiation fog than are hilltops.

______ 10. The largest concentration of condensation nuclei is usually observed at cloud level.

______ 11. On a winter night the air cools to the dew-point temperature and a thick layer of radiation fog forms around midnight. If the air continues to cool during the night, in 5 hours or so the dew-point temperature will probably be higher than it was at midnight.

______ 12. With the same water vapor content, fog that forms in clear air is usually thicker (more opaque) than fog that forms in dirty air.

______ 13. Dry haze usually restricts visibility more than wet haze.

PROBLEMS

1. Table 1 provides the air temperature and dew-point temperature for five cities at 5 PM. Also given is the minimum temperature for the following morning. If the sky remains clear during the night, in the space provided at the bottom of the table, write in whether the city should expect dew, frozen dew, or visible frost during the morning. If none of these is expected, write in the word "none".

Table 1

	CITY 1	CITY 2	CITY 3	CITY 4	CITY 5
Air Temperature,°F (5 PM)	50°	52°	46°	48°	57°
Dew Point,°F (5 PM)	40°	30°	30°	35°	21°
Minimum Temperature,°F (6 AM)	36°	28°	34°	32°	24°
Expect to observe					

2. In the space provided, write in the name of the cloud that is depicted in the illustration. (Additional photographs of clouds are found in the color cloud chart bound at the back of your textbook.)

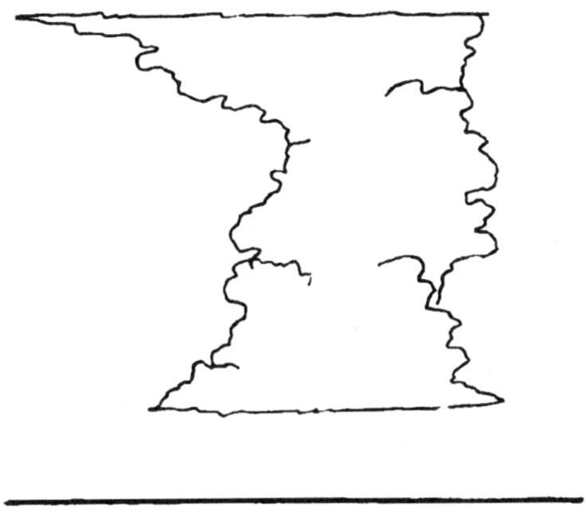

a. ______________________

Figure a

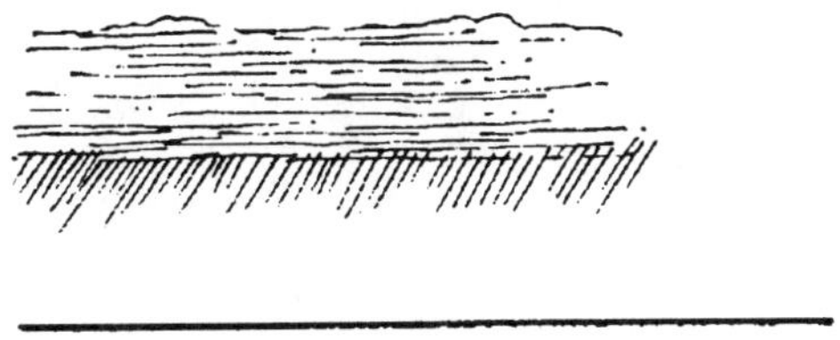

b. ______________________

Figure b

c. ______________________

Figure c

d. ______________________________

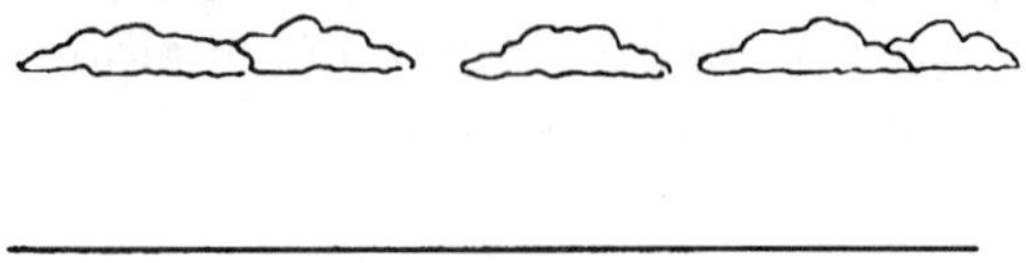

Figure d

e. ______________________________

Figure e

f. ______________________________

Figure f

g. ______________________________

Figure g

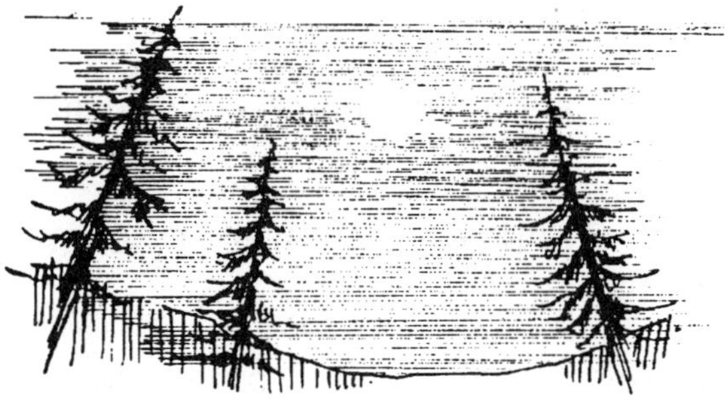

(middle cloud)

Figure h

h. ______________________________

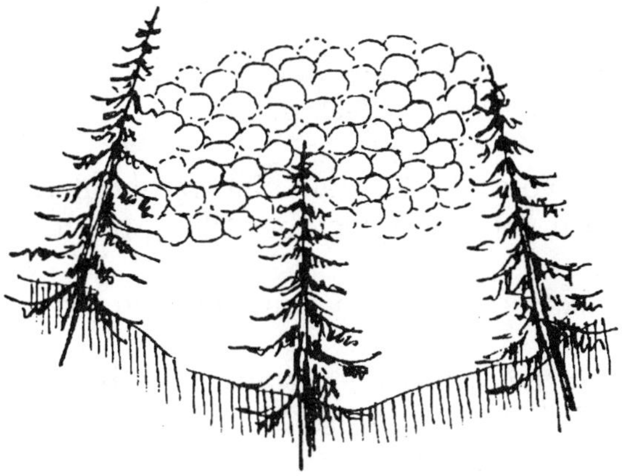

(middle cloud)

Figure i

i. ______________________________

(high cloud)

Figure j

j. ______________________________

ADDITIONAL READINGS

"Fog on Trial" by Stanley David Gedzelman, *Weatherwise*, Vol. 44, No. 2 (April 1991), p. 14.

"Five Faces of Freezing" by Craig F. Bohren, *Weatherwise*, Vol. 42, No. 6 (December 1989), p. 315.

"In Praise of Altocumulus" by Stanley David Gedzelman, *Weatherwise*, Vol. 41, No. 2 (June 1988), p. 143.

"The Mystery of Arctic Haze" by John Carey, *Weatherwise*, Vol. 41, No. 2 (April 1988), p. 97.

"Weather Satellites — Fulfilling the Promises" by Robert G. Cowen, *Weatherwise*, Vol. 37, No. 2 (April 1984), p. 64.

"What Your Windshield Shows about the Clouds" by Richard Williams, *Weatherwise*, Vol. 40, No. 5 (October 1987), p. 251.

"An Essay on Dew" by Craig Bohren, *Weatherwise*, Vol. 41, No. 4 (August 1988), p. 226.

A Field Guide to the Atmosphere by Vincent J. Schaefer and John A. Day; Houghton Mifflin, Boston, 1981.

"The Meteorological Satellite: Overview of 25 Years of Operation" by W. L. Smith, et. al., *Science*, Vol. 231, No. 4737 (January 31, 1986), p. 455.

"Alaska's Unusual Silver Clouds" by Robert E. Fischer, *Weatherwise*, Vol . 33, No. 5 (October 1980), p. 222.

"Controlling Fog" by Bruce A. Kunkel, *Weatherwise*, Vol. 33, No. 3 (June 1980), p. 117.

"All that Glistens Isn't Dew" by Craig F. Bohren, *Weatherwise*, Vol. 43, No. 5 (October 1990), p. 284.

"Cloud Classification Before Luke Howard" by Stanley David Gedzelman, *Bulletin of the American Meteorological Society*, Vol. 70, No. 4 (April 1989), p. 381.

"John Aitken's Contribution to Atmospheric and Aerosol Science — One Hundred Years of Condensation Nuclei Counting" by Josef Podzimek, *Bulletin of the American Meteorological Society*, Vol. 70, No. 12 (December 1989), p. 1538.

ANSWERS

MATCHING

1. g
2. c
3. a
4. l
5. i
6. b
7. e
8. k
9. f
10. h
11. j
12. d

ADDITIONAL MATCHING (CLOUDS)

1. c
2. h
3. j
4. k
5. d
6. b
7. a
8. f
9. e
10. k
11. i
12. e
13. c
14. g
15. h
16. g
17. k

FILL IN THE BLANK

1. ice crystals
2. advection
3. radiational
4. mammatus
5. scattered
6. contrail or condensation trail
7. hygroscopic
8. lenticular
9. noctilucent cloud
10. pileus or cap cloud
11. nacreous

MULTIPLE CHOICE

1. d
2. e
3. b
4. a
5. c
6. d
7. b
8. c
9. a
10. b
11. a
12. e
13. b
14. e
15. c

TRUE - FALSE

1. T
2. T
3. T
4. F
5. T
6. F
7. F
8. T
9. T
10. F
11. F
12. T
13. F

PROBLEMS

1. City 1, dew
   City 2, frost
   City 3, none
   City 4, dew and frozen dew
   City 5, none
2. a. cumulonimbus
   b. nimbostratus
   c. lenticular
   d. stratocumulus
   e. cirrostratus
   f. cumulus
   g. cirrus
   h. altostratus
   i. altocumulus
   j. cirrocumulus

CHAPTER

8

STABILITY AND CLOUD DEVELOPMENT

Chapter Eight ties together the concepts of stability and the formation of clouds. The first part of the chapter deals with atmospheric stability. Here we learn that stable air tends to resist upward vertical motions and that clouds forming in a stable atmosphere tend to spread horizontally and have a stratified appearance. Unstable air tends to enhance upward motions, causing clouds to develop vertically and build to great heights. After describing the causes of stable and unstable atmospheres, the chapter discusses the various ways clouds form. At this point we learn that instability generated by surface heating can produce cumulus clouds, and that topographic barriers can greatly influence cloud development. The latter part of the chapter describes how changes in atmospheric stability can promote changes in existing clouds, and how vertical mixing can change a clear day into a cloudy one.

Some important concepts of this chapter:

1. A parcel (blob) of air in stable equilibrium will tend to return to its original position, while a parcel (blob) of air in unstable equilibrium will tend to move away from its original position.
2. A rising parcel of *unsaturated* air will cool at the dry adiabatic rate, which is about 10°C per 1000 m (5.5°F per 1000 ft); similarly, a descending parcel of unsaturated air will warm at the same rate.
3. A rising parcel of *saturated* air will cool at the moist adiabatic rate, which varies, but an average of 6°C per 1000 m (3.3°F per 1000 ft) is often used; similarly, a descending parcel of saturated air will warm at the same rate.
4. The environmental lapse rate is the rate at which the actual air temperature decreases with increasing altitude above the surface.
5. The atmosphere is absolutely *stable* when the air at the surface is either cooler than the air aloft (an inversion), or the temperature difference between the warmer surface air and the air aloft is not very great, i.e., the environmental lapse rate is less than the moist adiabatic rate.
6. The atmosphere can be made more stable by cooling the surface air, warming the air aloft, or by causing air to sink (subside) over a vast area.
7. The atmosphere is absolutely *unstable* when the surface air is much warmer than the air aloft, i.e., the environmental lapse rate is greater than the dry adiabatic rate.
8. The atmosphere can be made more unstable by warming the surface air, cooling the air aloft, or by lifting a layer of air.

9. The atmosphere is conditionally unstable when unsaturated air can be lifted to a point where condensation occurs and the rising air becomes warmer than the air around it. This takes place when the environmental lapse rate lies between the moist adiabatic rate and the dry adiabatic rate.
10. The majority of clouds form due to surface heating, the convergence of surface air, and forced uplift along topographic barriers and weather fronts.

SELF TESTS

MATCH THE FOLLOWING

_____ 1. Type of uplift that occurs as air is lifted over a mountain barrier

_____ 2. A cloud with lightning and thunder that forms when the atmosphere is unstable

_____ 3. The mixing of outside air into a cloud or rising parcel of air

_____ 4. A vertical profile of air temperature

_____ 5. A measured increase in air temperature with increasing height

_____ 6. A low cloud that forms when the atmosphere is absolutely stable

_____ 7. Drier region observed on the downwind (leeward) side of a mountain range

_____ 8. This term is added to altocumulus or cirrocumulus when they have towerlike extensions

_____ 9. A cloud associated with turbulent eddies, downwind of a mountain

_____ 10. Wave clouds that may form over or downwind of a mountain

_____ 11. The process by which a parcel expands and cools or compresses and warms with no interchange of heat with its surroundings

_____ 12. This marks the base of a cloud that has formed by lifting

a. stratus
b. LCL
c. sounding
d. rotor
e. cumulonimbus
f. inversion
g. entrainment
h. castellanus
i. adiabatic
j. orographic
k. lenticular
l. rain shadow

■ FILL IN THE BLANK

1. If an air parcel is given a small push upward and it continues to move upward on its own accord, the atmosphere is said to be __________.

2. A __________ inversion forms when air slowly sinks over a wide area.

3. The rate at which the air temperature changes inside a rising or descending parcel of *unsaturated* air is called the __________ __________ __________.

4. If an air parcel is given a small push upward and it falls back to its original position, the atmosphere is said to be __________.

5. The rate at which the temperature changes inside a rising or descending parcel of *saturated* air is called the __________ __________ __________.

6. If unsaturated stable air is lifted to a level where it becomes saturated and unstable, this type of instability is called __________ __________.

■ MULTIPLE CHOICE

1. Rising saturated air cools at a lesser rate than rising unsaturated air primarily because:
   a. rising saturated air is heavier
   b. rising saturated air is lighter
   c. unsaturated air expands more rapidly
   d. saturated air does not expand
   e. latent heat of condensation is released in a rising parcel of saturated air

2. When the air temperature decreases very slowly with height, and the environmental lapse rate is *less* than the moist adiabatic rate, the atmosphere is:
   a. absolutely unstable
   b. absolutely stable
   c. conditionally unstable
   d. neutrally stable
   e. absolutely unhappy

3. When the atmosphere is absolutely unstable, the environmental lapse rate is __________ the dry adiabatic rate.
   a. equal to
   b. greater than
   c. less than

4. A moist layer of stable surface air can change into a deck of low stratocumulus clouds by the process of:
   a. lifting
   b. cooling the surface
   c. subsidence
   d. mixing
   e. warming the top of the moist layer

5. Clouds that appear as waves breaking along the shore are called:
   a. billows
   b. pileus
   c. mammatus
   d. rotors
   e. castellanus

6. For least polluted surface air, the best time of the day for a farmer to burn agricultural debris would be:
   a. around sunrise
   b. around sunset
   c. around the time of maximum surface air temperature
   d. about midnight, so no one will see it

7. These two conditions, working together, will make the atmosphere the most *unstable*:
   a. cool the surface and cool the air aloft
   b. cool the surface and warm the air aloft
   c. warm the surface and cool the air aloft
   d. warm the surface and warm the air aloft

8. These two conditions, working together, will make the atmosphere the most *stable*:
   a. cool the surface and cool the air aloft
   b. cool the surface and warm the air aloft
   c. warm the surface and cool the air aloft
   d. warm the surface and warm the air aloft

9. Which of the following is *not* a way of producing clouds?
   a. lifting air over a topographic barrier
   b. lifting air along a weather front
   c. warming the surface of the earth
   d. convergence of surface air
   e. subsidence

10. An inversion represents an extremely stable atmosphere because a parcel of air that rises into an inversion will eventually become __________ and __________ dense than the air surrounding it.

a. warmer, more
b. warmer, less
c. colder, more
d. colder, less

11. Subsidence inversions are best developed with __________ pressure areas because of the __________ air associated with them.

a. high, sinking
b. high, rising
c. low, sinking
d. low, rising

12. The moist adiabatic rate is different from the dry adiabatic rate because:

a. saturated air is always unstable
b. an unstable air parcel expands more rapidly
c. unsaturated air is always stable
d. latent heat is released inside a parcel of rising saturated air
e. a parcel of saturated air weighs less than a parcel of unsaturated air

■ TRUE - FALSE

______ 1. Convection can take place over water.

______ 2. Most thunderstorms do not penetrate very far into the stratosphere because the stratosphere is a layer of unstable air.

______ 3. Air motions tend to be downward around the outside of a cumulus cloud.

______ 4. A conditionally unstable atmosphere is stable with respect to unsaturated air and unstable with respect to saturated air.

______ 5. Radiational cooling of the surface at night tends to make the lower atmosphere unstable.

______ 6. When cloud elements become arranged in rows they are called cloud streets.

______ 7. The smoke plume that often raises the concentration of surface pollutants to dangerously high levels is fumigation.

______ 8. In a conditionally unstable atmosphere, the environmental lapse rate will be greater than the moist adiabatic rate and less than the dry adiabatic rate.

______ 9. Lowering an entire layer of air will generally make it more unstable.

______ 10. Warming the air aloft will tend to make the atmosphere more unstable.

______ 11. If you were to take a trip this summer from Ohio to Nevada, you would probably observe the bases of afternoon cumulus clouds increasing in height above the ground as you travel westward.

______ 12. Normally, during the course of a day, the atmosphere near the surface is the most unstable during the afternoon, and the most stable in the early morning.

■ PROBLEMS

1. Suppose the environmental lapse rate is 4°F per 1000 feet. If the air temperature at the surface (0 feet) is 70°F, then the air temperature measured by a radiosonde at 5000 feet above the surface should be __________ °F.

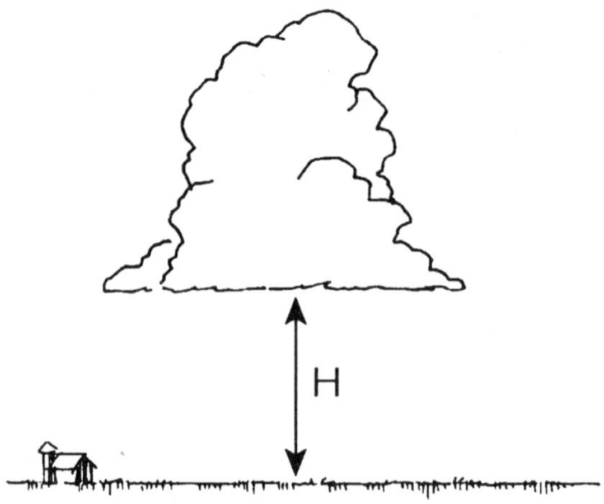

Figure 1

2. a. In Figure 1, what would be the approximate height of the base of the cumulus cloud (H), when the surface air temperature is 86°F, and the surface dew-point temperature is 68°F? (Hint: use the formula given on p. 204 in your text.)

b. Suppose the cloud in Figure 1 builds to a height of 14,000 feet above the surface. What would be the approximate temperature, inside the cloud, at an altitude of 14,000 ft? ((Hint: use 5.5°F/1000 ft for the dry adiabatic rate and 3.3°F/1000 ft for the moist adiabatic rate.)

c. If the temperature of the air surrounding the cloud at 14,000 ft is 34°F, would the cloud continue to build vertically? Explain.

d. Since this cumulus cloud has developed into a much larger cloud, it would now be called a ______________________.

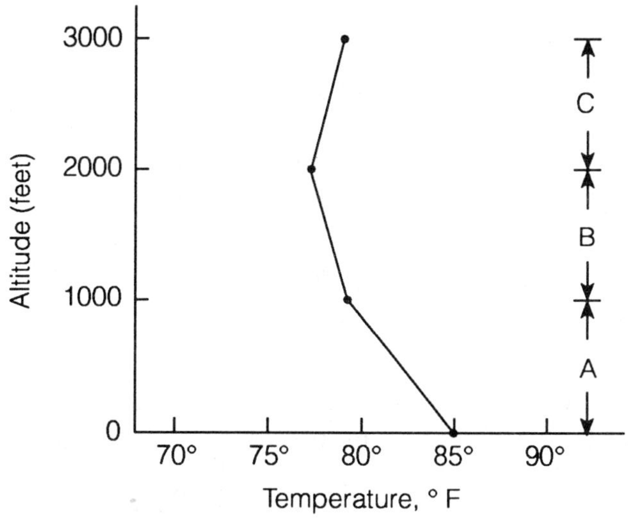

Figure 2

3. Figure 2 represents a sounding — a vertical profile of air temperature as measured by a radiosonde.

a. Which of the three layers (A, B or C) is the most unstable? __________

b. Which of the three layers is the most stable? __________

c. In which of the three layers is there an inversion? __________

d. If the air temperature at the bottom of layer A is 85°F, and if the air temperature at the top of layer C is 79°F, what would be the average environmental lapse rate from the surface to 3000 ft? __________

e. Even though the bottom layer (A) may be unstable, the lapse rate you calculated in (d) indicates that the entire region from the surface to 3000 ft represents what type of atmospheric stability? __________

f. If a cumulus humilis cloud formed at a level of about 1000 ft, the top of the cloud would probably not be much higher than what elevation? __________

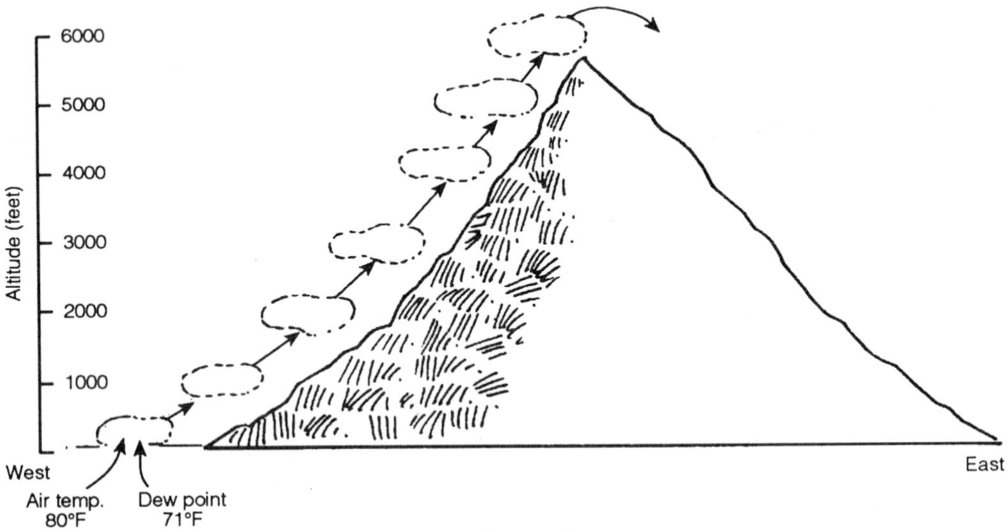

Figure 3

4. In Figure 3, suppose a parcel of air with a temperature of 80°F and a dew-point temperature of 71°F is lifted up and over a 6000-foot mountain.

a. As the parcel rises (as indicated by the blobs), at approximately what elevation would condensation begin and a cloud start to form? __________ (Hint: Use the dry adiabatic rate of 5.5°F/1000 ft and remember for unsaturated air the dew-point temperature in a rising parcel decreases by about 1°F/1000 ft.)

b. What is the name given to this condensation level? ________ ________ ________

c. What is the air temperature and dew-point temperature at the condensation level?
__________ °F

d. What is the relative humidity of the rising parcel at the condensation level?
__________ %

From the condensation level a cloud forms and the parcel continues its upward journey. Keep in mind that now the air temperature and the dew-point temperature both decrease at the same rate — the moist adiabatic rate. (To simplify matters, use a moist adiabatic rate of 3.0°F/1000 ft.)

e. At the top of the mountain, inside the cloud, what is the air temperature and the dew-point temperature of the rising parcel of air? __________ °F

Suppose the cloud builds only to the top of the mountain and does not lap over to the other side. Further suppose that the air parcel at the top of the mountain with the temperature and dew point calculated in (e) descends the eastern side all the way to the base of the mountain (0 ft). (Hint: Remember that the air is now unsaturated and that the air temperature will increase at the dry adiabatic rate of 5.5°F/1000 ft and the dew-point temperature of the descending air will increase by about 1°F/1000 ft.)

f. What would be the air temperature of the descending parcel at 0 feet on the eastern side of the mountain? __________ °F

g. How many degrees warmer is the sinking air on the eastern side of the mountain at 0 feet compared to the air at the same level on the western side? __________ °F

h. What accounts for the air being warmer?

i. What would be the dew-point temperature of the descending parcel at 0 feet on the eastern side of the mountain? __________ °F

j. The lower dew-point temperature on the eastern side of the mountain represents drier air. What accounts for this lower dew point?

5. A few adiabatic chart questions. The following questions can be answered by using the adiabatic chart (Figure 7) on p. 207 of your textbook.

a. Suppose a parcel of surface air at 0 meters (pressure 1013 mb) has a temperature of 10°C. If this air parcel is lifted at the dry adiabatic rate to a level where the pressure is 700 mb (about 3 km), what would be the temperature inside the parcel at this level? __________ °C

b. What would be the saturation mixing ratio (w_s) of the air in the parcel when its temperature is 10°C and it is at the surface (0 meters, pressure 1013 mb)? __________ g/kg

c. If the dew-point temperature inside this parcel (0 meters, pressure 1013 mb) is 1.0°C, what would be its actual mixing ratio? __________ g/kg

d. With the saturation mixing ratio (w_s) you obtained in (b) and the actual mixing ratio you obtained in (c), compute the relative humidity inside the air parcel. __________% (Hint: an example is given on p. 206-207 in your text.)

ADDITIONAL READING

"Smoke Signals" by Alfred Blackadar, *Weatherwise*, Vol, 41, No. 2 (June 1988), p. 159.

"Altocumulus Lenticularis Clouds over Flat Terrain" by E. S. S. Takle and John M. Brown, *Weatherwise*, Vol. 35, No. 3 (June 1982), p. 131.

A Field Guide to the Atmosphere by Vincent J. Schaefer and John A. Day; Houghton Mifflin, Boston, 1981. See especially pp. 65-73.

"The Art and Physics of Soaring" by Lloyd Hunter, *Physics Today*, Vol. 37, No. 4 (April 1984), p. 34.

ANSWERS

MATCHING

1.	j	4.	c	7.	l	10.	k
2.	e	5.	f	8.	h	11.	i
3.	g	6.	a	9.	d	12.	b

FILL IN THE BLANK

1. unstable
2. subsidence inversion
3. dry adiabatic rate
4. stable
5. moist adiabatic rate
6. conditional instability

MULTIPLE CHOICE

1.	e	4.	d	7.	c	10.	c
2.	b	5.	a	8.	b	11.	a
3.	b	6.	c	9.	e	12.	d

TRUE - FALSE

1.	T	4.	T	7.	T	10.	F
2.	F	5.	F	8.	T	11.	T
3.	T	6.	T	9.	F	12.	T

PROBLEMS

1. 50°F
2. a. 3996' or approximately 4000'
   b. 31°F
   c. The cloud would not continue to build vertically because the rising air inside the cloud is colder and, therefore, more dense (heavier) than the air surrounding it.
   d. cumulus congestus
3. a. layer A
   b. layer C
   c. layer C
   d. 2°F/1000 ft
   e. stable or absolutely stable
   f. 2000 ft
4. a. approximately 2000 ft
   b. Lifting Condensation Level (LCL)
   c. 69°F
   d. 100%
   e. 57°F
   f. 90°F
   g. 10°F
   h. On the western side of the mountain, latent heat of condensation was released as sensible heat in the rising saturated air. This caused the temperature inside the cloud at the top of the mountain to be warmer than it would have been had the cloud not formed. Along with this, on the eastern side, the sinking air was compressed and warmed at the dry adiabatic rate from the top of the mountain down to its base.
   i. 63°F
   j. On the western side of the mountain, water vapor was removed from the air during condensation and probably remained on the western side as precipitation.
5. a. about -20°C
   b. about 9 g/kg
   c. about 4 g/kg
   d. about 44%

CHAPTER

9

PRECIPITATION

The beginning of Chapter Nine describes how cloud droplets are able to grow large enough to fall as rain or snow. This is followed by a section that explains how falling raindrops and snowflakes can be changed into other forms of precipitation, such as sleet and freezing rain. Here we learn about the different types of precipitation and how they influence our environment by affecting how far we can see and hear. The last section describes the various methods used in measuring precipitation.

Some important concepts of this chapter:

1. Cloud droplets are much smaller than raindrops.
2. The curvature effect tells us that tiny cloud droplets evaporate more quickly than larger droplets.
3. Hygroscopic ("water-seeking") particles dissolved in pure water reduce the relative humidity necessary for the onset of condensation, so that condensation may begin on certain particles when the relative humidity is considerably less than 100 percent. This is called the solute effect.
4. The combining (merging) of tiny cloud droplets into a larger droplet is called coalescence.
5. During the ice crystal (Bergeron) process, ice crystals, surrounded by water droplets, grow larger at the expense of the droplets; consequently as ice crystals grow larger, water droplets become smaller.
6. Accretion is the process by which ice crystals or snowflakes grow larger by colliding with supercooled liquid droplets that freeze on contact.
7. It is never too cold to snow.
8. Snow can fall when the air temperature is considerably above freezing.
9. Freezing rain forms when cold raindrops fall through a shallow, subfreezing layer and freeze upon striking the ground or the surface of a cold object.
10. Sleet forms when cold raindrops, or partially melted snowflakes, fall through a relatively deep subfreezing layer and turn back into solid ice before reaching the surface.
11. Hailstones only form in thunderstorms that have updrafts capable of keeping ice particles within the cloud long enough to acquire further coatings of ice.

SELF TESTS

MATCHING

______	1. An aggregate of ice crystals	a.	snow squall
______	2. High winds, low temperatures and blowing or falling snow	b.	Bergeron process
______	3. Radio detection and ranging	c.	blizzard
______	4. Injecting a cloud with small particles with the intent to enhance precipitation	d.	hailstreak
______	5. An intense snow shower	e.	shower
______	6. The smallest raindrops	f.	fall streaks
______	7. The largest form of precipitation	g.	radar
______	8. Intermittent, heavy downpour of either rain or snow	h.	drizzle
______	9. The most common snowflake shape	i.	snowflake
______	10. A cloud is said to be this when only ice crystals exist in it	j.	glaciated
______	11. Particles in the atmosphere on which ice crystals may grow	k.	hail
______	12. Another name for the ice-crystal process of rain formation	l.	seeding
______	13. These form when snowflakes or ice crystals fall from high cirriform clouds	m.	flurries
______	14. A deposit of hail in a long narrow band	n.	ice nuclei
______	15. A light shower of snow that falls intermittently from cumuliform clouds for a short duration	o.	dendrite

FILL IN THE BLANK

1. The merging of cloud droplets by collision is called _______________.

2. An amount of precipitation measured to be less than one hundredth of an inch is called a _______________.

3. The two main substances used in cloud seeding are ____________ and ____________.

4. The growth of a precipitation particle by the collision of an ice crystal or snowflake with a liquid droplet at temperatures below freezing is called _______________.

5. A cold raindrop (or partially melted snowflake) that freezes into a pellet of ice in a deep, subfreezing layer of surface air is called _______________.

6. Water droplets that exist at temperatures below freezing are said to be _______________.

7. Cold rain that falls through a rather shallow layer of subfreezing air and freezes upon striking a surface is called _______________ _______________.

8. Rain that falls from a cloud but evaporates before reaching the surface is called _______________.

MULTIPLE CHOICE

1. Brittle, crunchy pieces of snowlike ice that usually fall as a shower from a cumuliform cloud are:

   a. sleet
   b. freezing rain
   c. snow grains
   d. snow pellets
   e. hail

2. When the atmospheric relative humidity is below 100 percent, liquid cloud droplets may grow larger by condensation because of the:

   a. solute effect
   b. water effect
   c. curvature effect
   d. hydrophobic effect
   e. adiabatic effect

3. Which of the adjacent illustrations best describes the ice-crystal process of rain formation?

   a. 1
   b. 2
   c. 3
   d. 4

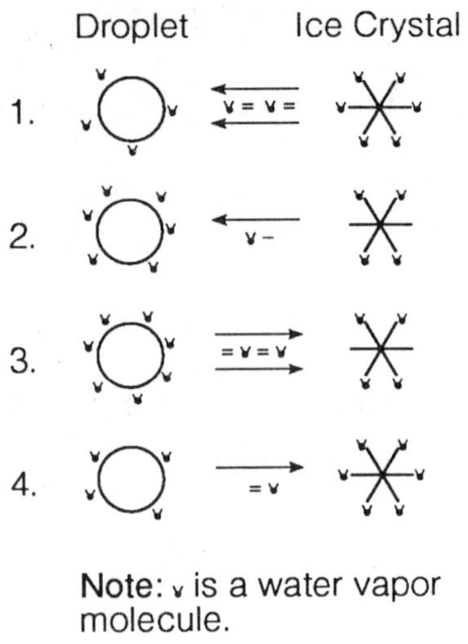

4. Of the clouds listed below, which would most likely produce drizzle?

   a. cumulus
   b. stratus
   c. altostratus
   d. cumulus congestus
   e. cumulonimbus

5. If it is raining on one side of the street but not on the other, it is a good bet that the rain is falling from:

   a. a nimbostratus cloud
   b. an altocumulus cloud
   c. a stratus cloud
   d. an altostratus cloud
   e. a cumulonimbus cloud

6. Which of the following is *not* considered an important factor in the production of rain by the collision-coalescence process?

   a. the updrafts in the cloud
   b. the number of ice crystals in the cloud
   c. the cloud thickness
   d. the relative size of the droplets
   e. the cloud's liquid water content

7. Which below best describes the shape of a large falling raindrop about 5 mm (5000 micrometers) in diameter?

   a. spherical
   b. tear drop
   c. square
   d. elongated like that of a cylinder
   e. slightly elongated and flattened on the bottom

8. Which of the following conditions would be most suitable for natural cloud seeding?

   a. cirrus clouds above altostratus clouds
   b. altocumulus clouds above stratus clouds
   c. stratus clouds above ground fog
   d. cirrostratus clouds above stratocumulus clouds
   e. altostratus clouds above stratus clouds

9. Lumpy ice particles, also known as snow pellets, that form in a cloud usually by the process of accretion are called:

   a. sleet
   b. glaze
   c. graupel
   d. snow grains
   e. snow flakes

10. Hail forms inside this cloud:

   a. castellanus
   b. cumulonimbus
   c. stratocumulus
   d. cumulus humilis
   e. lenticular

11. The freezing of supercooled droplets by contact with a nucleus is called:

   a. coalescence
   b. riming
   c. spontaneous nucleation
   d. contact nucleation

■ TRUE - FALSE

______ 1. It is never too cold to snow.

______ 2. Ice nuclei are more abundant than condensation nuclei.

______ 3. You would use a wooden stick to measure rainfall in the tipping bucket rain gauge.

______ 4. Inside a cloud, when the relative humidity is 100 percent, larger cloud droplets evaporate more quickly than do the smaller cloud droplets.

______ 5. Snow can reach the surface when the surface air temperature is considerably above freezing as long as the surface air is not saturated.

_____ 6. Radar gathers information about precipitation in clouds by measuring the amount of sunlight reflected off the precipitation.

_____ 7. Much of the rain that falls in middle and high latitudes of North America begins as snow.

_____ 8. Generally, the smaller the amount of pure water, the lower the temperature at which the water will freeze.

_____ 9. Large, heavy snowflakes are generally associated with moist air and temperatures well below freezing.

_____ 10. Cloud seeding using silver iodide only works in cold clouds.

_____ 11. During the ice-crystal process of rain formation, ice crystals grow at the expense of the surrounding water droplets.

_____ 12. In a cloud, small droplets tend to fall faster than large droplets.

_____ 13. It takes about one million average size cloud droplets to produce an average size raindrop.

_____ 14. Glaze is another name for sleet.

_____ 15. On the wing of an aircraft, rime ice is usually more difficult to remove than clear ice.

_____ 16. Freshly fallen snow will usually produce a creaking sound when the air is cold, usually below -10°C (14°F).

■ PROBLEMS

1. After a large snowstorm the newspaper reports the water equivalent of the precipitation for the following cities to be:

			snowfall
a.	Albany, NY	1.35 inches	_____
b.	Chicago, IL	0.83 inches	_____
c.	Cleveland, OH	0.91 inches	_____
d.	Detroit, MI	0.25 inches	_____

If we assume an average water equivalent ratio of 1:10 for this storm, and that all of the precipitation fell as snow, then how many inches of snow did each city receive?

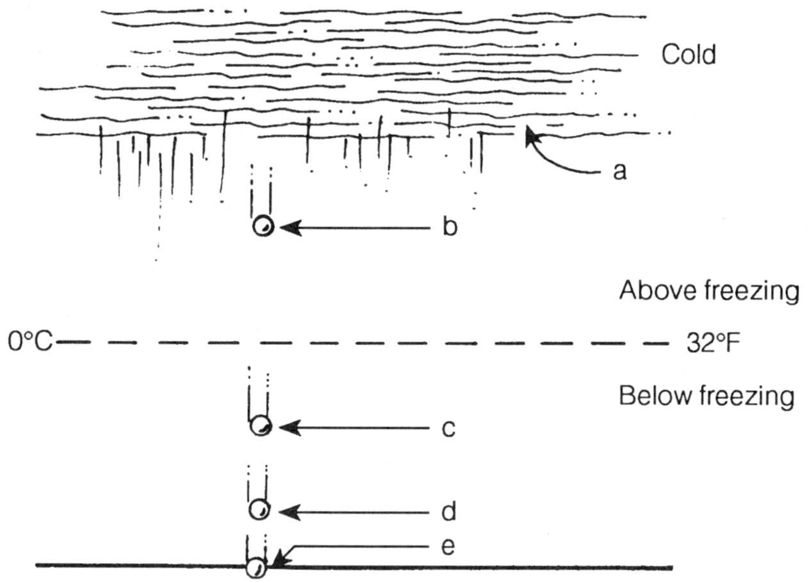

Figure 1

2. Answer the following questions that pertain to letters a through e in Figure 1.

   a. The thick, relatively low cloud in Figure 1 is stratified and producing light precipitation over a broad area. What is the name of this cloud? __________

   b. At letter b, a cold raindrop is falling toward the surface in above-freezing air. If the raindrop initially formed in the cold upper part of the cloud as a snowflake, the precipitation probably formed by the __________ process of rain formation.

   c. The cold, liquid raindrop at letter c, falling through this region of below-freezing air, is said to be ____________________.

   d. If the cold raindrop freezes at letter d before reaching the ground it will produce precipitation called __________.

   e. If the cold raindrop freezes upon striking the surface at letter e, the precipitation is called __________ __________.

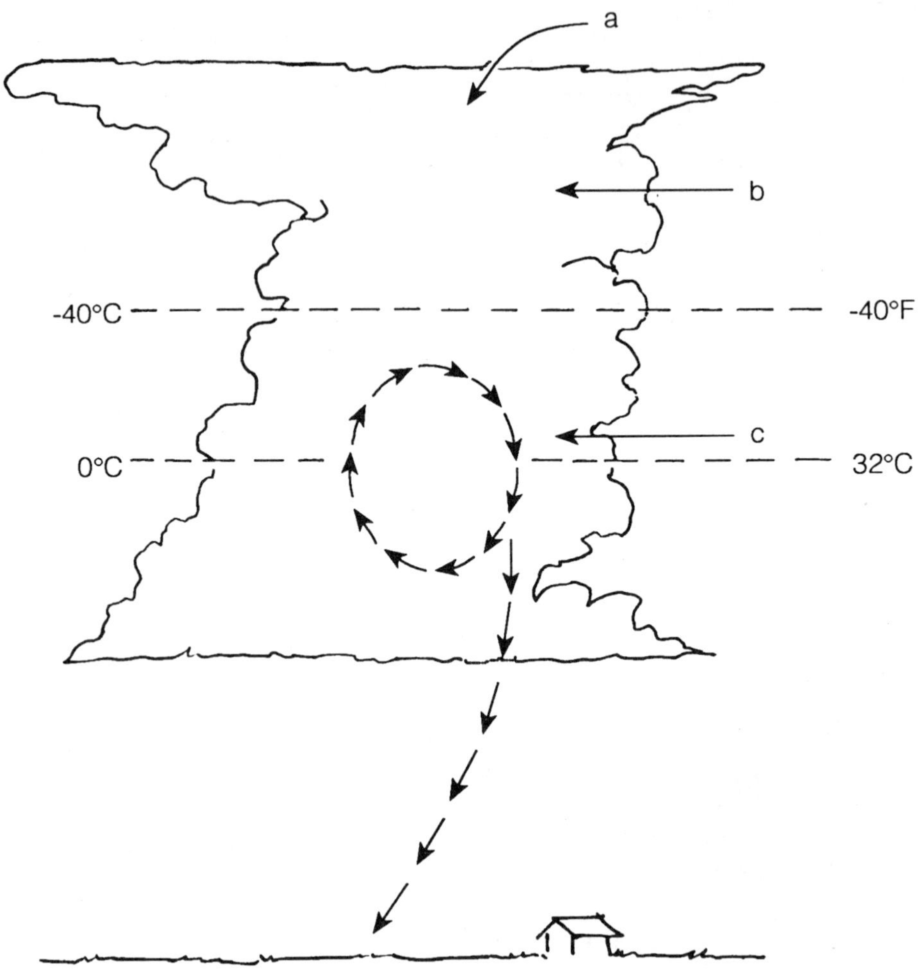

Figure 2

3. Answer the following questions that pertain to Figure 2.

   a. The cloud in Figure 2 is what type? ______________

   b. What particles would you most likely expect to observe in the region of letter a and letter b in this cloud? ____________________

   c. What particles would you most likely expect to observe in the region of letter c? ____________________

   d. Suppose a precipitation particle made several up and down journeys, as indicated by the arrows. If the precipitation eventually reaches the surface as ice, it would be called ______________.

ADDITIONAL READINGS

"How Snow Crystals Grow" by John Hallett, *American Scientist*, Vol. 72, No. 6 (Nov-Dec 1984), p, 582.

"Bentley and Lenard: Pioneers in Cloud Physics" by Duncan C. Blanchard, *American Scientist*, Vol. 60, No. 6 (Nov-Dec 1972), p. 746.

"Holes in Clouds?" by Peter V. Hobbs, *Weatherwise*, Vol. 38, No. 5 (October 1985), p. 254.

"What Becomes of a Winter Snowflake? " by Samuel C . Colbeck, *Weatherwise*, Vol. 38, No. 6 (December 1985), p. 312.

"Of Wet Snow, Slush and Snowballs" by Samuel C. Colbeck, *Weatherwise*, Vol. 39, No. 6 (December 1986), p. 314.

"A Memorable Easter Ice Storm" by Mary Reed, *Weatherwise*, Vol. 40, No. 2 (April 1987), p. 78.

Clouds in a Glass of Beer by Craig F. Bohren; John Wiley and Sons, Inc., New York, 1987. See especially pp. 1-14.

"Precipitation — The Forms and Effects of Ice and Water" by Vincent J. Schaefer and John A. Day, *Weatherwise*, Vol. 34, No. 6 (December 1981), p. 244.

"The Formation of Ice Crystals by Sublimation" by Vincent J. Schaefer, *Weatherwise*, Vol. 32, No. 6 (December 1979), p. 256.

"Methods of Forming Ice Crystals in a Supercooled Cloud" by Vincent J. Schaefer, *Weatherwise*, Vol, 32, No. 5 (October 1979), p. 217.

"Nuclei, Water Drops and Ice Crystals" by Vincent J. Schaefer, *Weatherwise*, Vol. 32, No. 4, (August 1979), p. 172.

"Oregon's Silver Thaw" by Fred W. Decker, *Weatherwise*, Vol. 32 , No. 2 (April 1979), p. 76.

"Hawaiian Hailstones — 30 January 1985" by Tsutomu Takahashi, *Bulletin of the American Meteorological Society*, Vol. 68, No. 12 (December 1987), p. 1530.

"Experimentation Involving Controversial Scientific and Technological Issues: Weather Modification as a Case Illustration" by Stanley A. Changnon and W. Henry Lambridge, *Bulletin of the American Meteorological Society*, Vol. 71, No. 3 (March 1990), p. 334.

ANSWERS

MATCHING

1. i
2. c
3. g
4. l
5. a
6. h
7. k
8. e
9. o
10. j
11. n
12. b
13. f
14. d
15. m

FILL IN THE BLANK

1. coalescence
2. trace
3. silver iodide and dry ice (solid carbon dioxide)
4. accretion
5. sleet
6. supercooled
7. freezing rain
8. virga

MULTIPLE CHOICE

1. d
2. a
3. c
4. b
5. e
6. b
7. e
8. a
9. c
10. b
11. d

TRUE - FALSE

1. T
2. F
3. F
4. F
5. T
6. F
7. T
8. T
9. F
10. T
11. T
12. F
13. T
14. F
15. F
16. T

PROBLEMS

1. a. 13.5 inches
   b. 8.3 inches
   c. 9.1 inches
   d. 2.5 inches
2. a. nimbostratus
   b. ice-crystal
   c. supercooled
   d. sleet (ice pellets)
   e. freezing rain (glaze)
3. a. cumulonimbus
   b. ice particles or ice crystals
   c. mostly liquid cloud droplets (water particles), some ice
   d. hail

CHAPTER

10

THE ATMOSPHERE IN MOTION — CHARTS, FORCES, AND WINDS

Chapter Ten gives us a broad view of how and why the wind blows. It opens with a section on surface and upper air charts. This is followed by a section that examines the forces that influence atmospheric motions. Here we learn that the wind blows in response to differences in atmospheric pressure and that once air begins to move, the Coriolis force tends to bend it to the right of its intended path in the Northern Hemisphere and to the left in the Southern Hemisphere. The latter part of the chapter deals with how the wind blows around pressure systems in the Northern and Southern Hemisphere, both aloft and at the surface.

Some important concepts of this chapter

1. Cold air aloft is normally associated with low pressure (low height), while warm air aloft is associated with high pressure (high height).
2. High and low pressures on a constant height chart correspond to high and low heights on a constant pressure (isobaric) chart.
3. The amount of pressure change that occurs over a given horizontal distance is the pressure gradient.
4. Horizontal differences in pressure create a pressure gradient force (PGF). This force causes the wind to blow.
5. On a weather map, closely spaced isobars (or contours) represent a steep pressure gradient, a strong PGF and high winds, while widely spaced isobars (or contours) represent a gentle pressure gradient, a weak PGF and light winds.
6. The Coriolis force causes the wind to bend to the right of its path in the Northern Hemisphere and to the left of its path in the Southern Hemisphere.
7. The Coriolis force only influences the direction of the wind and not the wind speed.
8. The Coriolis force increases with increasing latitude and with increasing wind speed.
9. The winds on an upper level chart tend to blow parallel to the isobars or contours in a more or less west-to-east direction in both hemispheres.
10. Surface winds tend to cross the isobars at an angle which averages about 30°. In the Northern Hemisphere, surface winds blow clockwise and outward from the center of a high, counterclockwise and inward toward the center of a low. In the Southern Hemisphere, surface winds blow counterclockwise and outward from the center of a high, clockwise and inward toward the center of a low.

SELF TESTS

MATCH THE FOLLOWING

_____ 1. The amount of pressure change that occurs over a given horizontal distance

_____ 2. Wind flow pattern that generally moves from west to east

_____ 3. Lines connecting points of equal elevation above sea level

_____ 4. An apparent force created by the rotation of the earth

_____ 5. Wind aloft that blows in a straight line at a constant speed parallel to the isobars or contours

_____ 6. The force which balances the vertical pressure gradient force and prevents the atmosphere around the earth from rushing off into space

_____ 7. Another name for a constant pressure chart

_____ 8. Wind flow pattern that takes on a more or less north-south trajectory

_____ 9. Wind aloft that blows at a constant speed parallel to curved isobars or contours

_____ 10. To correctly monitor horizontal changes in air pressure, this is the most important correction added to station pressure

a. meridional

b. geostrophic wind

c. contour lines

d. isobaric

e. altitude

f. pressure gradient

g. gradient wind

h. Coriolis

i. gravity

j. zonal

■ FILL IN THE BLANK

1. The force that causes the wind to deflect to the right of its path in the Northern Hemisphere and to the left in the Southern Hemisphere is the ________ ________.

2. Lines connecting points of equal pressure are called ____________.

3. The fundamental laws of motion were formulated by this man: ________ ________.

4. The force that initially causes the wind to blow is the ________ ________ ________.

5. On an upper-level chart (such as the 500 millibar chart), ________ lines are drawn to show horizontal changes in elevation.

6. Buys-Ballot's law states that, "In the Northern Hemisphere if you stand with your back to the surface wind, and turn clockwise about 30°, at the surface, lower pressure will be to your ________."

■ MULTIPLE CHOICE

1. The Coriolis force is strongest when the wind speed is ________ and the latitude is ________.

   a. high, low
   b. high, high
   c. low, high
   d. low, low

2. Air is in hydrostatic equilibrium when the upward-directed pressure force is balanced by the downward:

   a. Coriolis force
   b. pressure gradient force
   c. gradient force
   d. centripetal force
   e. force of gravity

3. The surface weather map is a sea level pressure chart. Thus, a surface weather map is also called:

   a. a constant pressure chart
   b. a constant height chart
   c. an isobaric chart

4. The surface winds around an area of low pressure normally _________. Above the system, the winds normally _________.

   a. diverge, converge
   b. diverge, diverge
   c. converge, diverge
   d. converge, converge

5. If, at your home in the Northern Hemisphere, the surface wind is blowing from the northeast, then according to Buys-Ballot's law, the region of lowest surface pressure will be to the _________ of your home.

   a. west
   b. north
   c. east
   d. south

6. Suppose a pilot flies from a region of cold air into a region of warm air. In the warm air the aircraft is indicating the same altitude as it did in the cold air, yet the aircraft would be flying:

   a. lower than the altimeter indicates
   b. higher than the altimeter indicates
   c. at the same elevation that the altimeter indicates

7. If an aircraft flies from standard temperatures into colder than standard air, without making any corrections, the altimeter in the colder air would indicate an altitude:

   a. higher than the aircraft's actual altitude
   b. lower than the aircraft's actual altitude
   c. exactly the same as the aircraft's altitude

8. In the Northern Hemisphere directly above you (at about 10,000 feet), clouds are moving from south to north, indicating a south wind. From Buys-Ballot's law you know that the center of lowest pressure aloft must be to the _________ of you.

   a. west
   b. east
   c. north
   d. south

9. On a weather map, the strongest winds are normally observed:

   a. at the center of high pressure
   b. at the center of low pressure
   c. near a large body of water
   d. where the isobars or contour lines are close together

■ TRUE - FALSE

_____ 1. The pressure gradient force is directed from high pressure toward lower pressure at all places on the earth.

_____ 2. To obtain *station pressure* with a mercury barometer, normally you must make a correction for altitude.

_____ 3. The Coriolis force causes the wind to blow faster.

_____ 4. Low pressures on a constant height chart correspond to low heights on a constant pressure chart.

_____ 5. The Coriolis force is the result of horizontal differences in pressure.

_____ 6. The winds aloft in middle latitudes of both hemispheres blow primarily from the west because the air aloft above high latitudes is colder than the air aloft above low latitudes.

_____ 7. If the earth stopped rotating, surface winds would blow directly from high pressure toward lower pressure.

_____ 8. On an *upper level chart*, winds tend to cross the isobars or contours at an angle that averages about 30°.

_____ 9. Cold air aloft is usually associated with low pressure and warm air aloft with high pressure.

_____ 10. The Coriolis force is zero at the equator.

_____ 11. As the wind moves in a curved path, the centripetal force results from an imbalance between the Coriolis force and the pressure gradient force.

_____ 12. If in the Northern Hemisphere at your home the surface wind is blowing from the southeast, then the region of lowest surface pressure will probably be to the east of your home.

_____ 13. Where the air aloft is warm, constant pressure surfaces are usually observed at a higher elevation than normal.

_____ 14. If, in the Northern Hemisphere at your home the surface wind is blowing from the southwest, then the region of lowest surface pressure will probably be to the south of your home.

______ 15. On a weather map, closely spaced isobars or contour lines indicate a region of high winds because in this region there is a strong pressure gradient force.

PROBLEMS

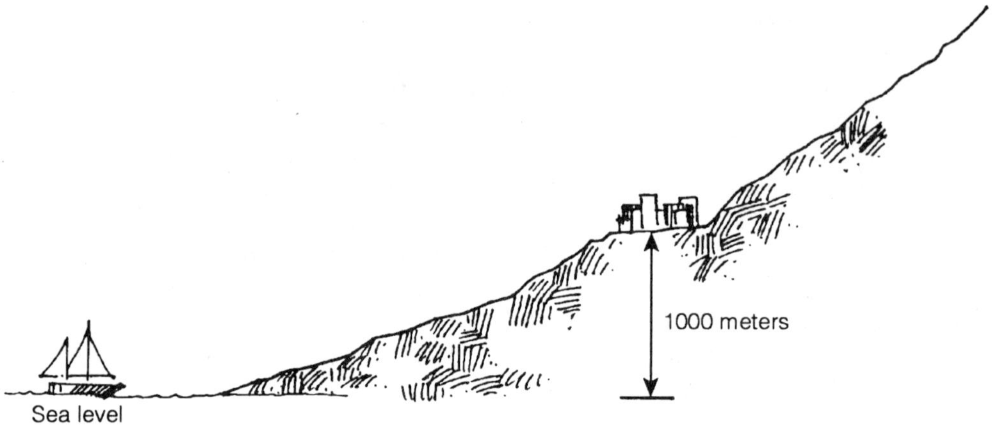

Figure 1

1. What is the approximate sea level pressure for the city in Figure 1 when its station pressure is 920 millibars? (Use a pressure change of 10 mb/100 m.)

   __________ mb

2. Each of the four illustrations in Figure 2 represent the wind-flow pattern around an area of high or low pressure in either the Northern or Southern Hemisphere.

   a. In the center of each pressure system indicate with the letter L or H whether it is a low or high pressure area.

   b. In the space provided beneath each pressure system, write in whether this system is located in the Northern Hemisphere or Southern Hemisphere.

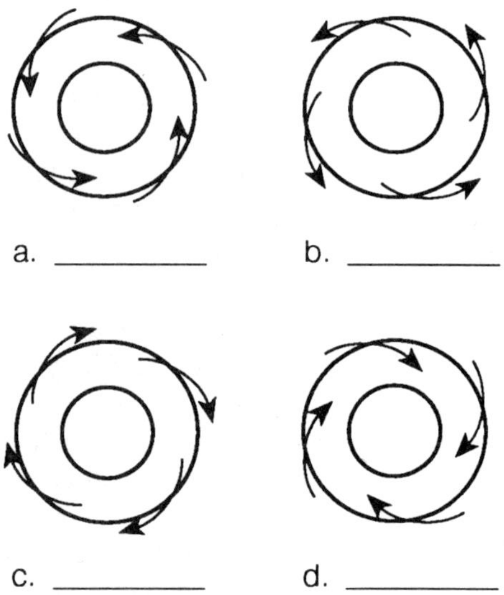

Figure 2

3. In Figure 3 show with arrows how the wind would blow around the area of low pressure and the area of high pressure on the upper-air chart. Both are located in the Northern Hemisphere and both are found aloft, above the level of friction. The lines around the L and H represent contour lines or isobars.

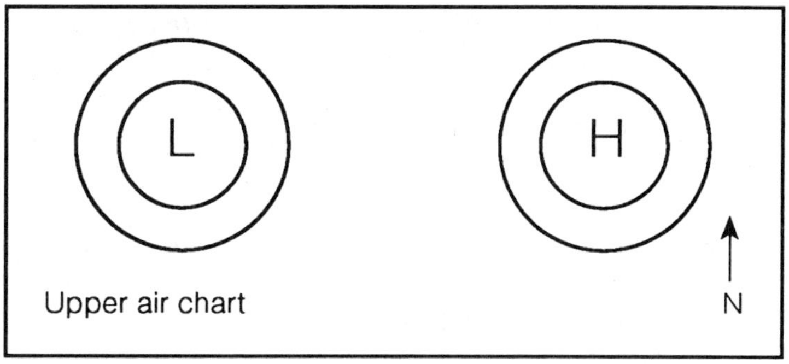

Figure 3

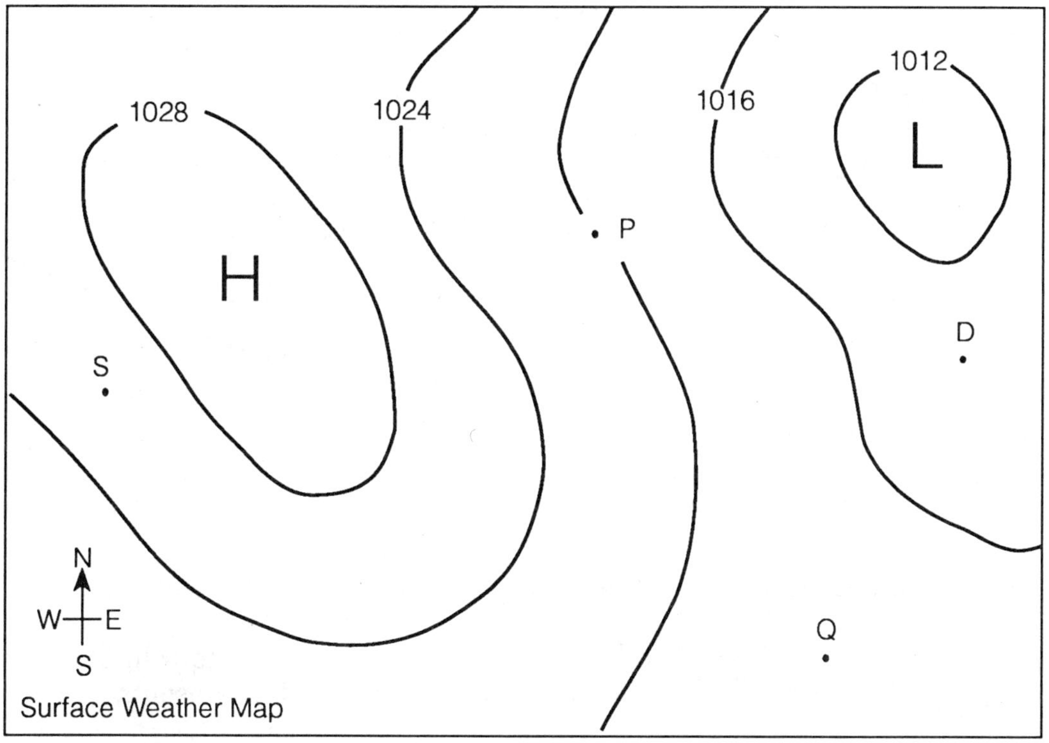

Figure 4

4. Answer the following questions that pertain to Figure 4, a surface weather map.

   a. What is the sea level pressure at point P? __________

   b. At point P, the wind would *most likely* be blowing from the __________.

   c. Would the pressure gradient force at point P be directed toward the H, the L, point D, or point Q? __________

d. Would the wind at point S *most likely* be blowing from the southeast, southwest, northeast, or northwest? __________

e. Would you expect the strongest winds at point P, point D or point Q? __________ Explain.

f. Should the air above the L be rising or sinking? __________

Should the air above the H be rising or sinking? __________

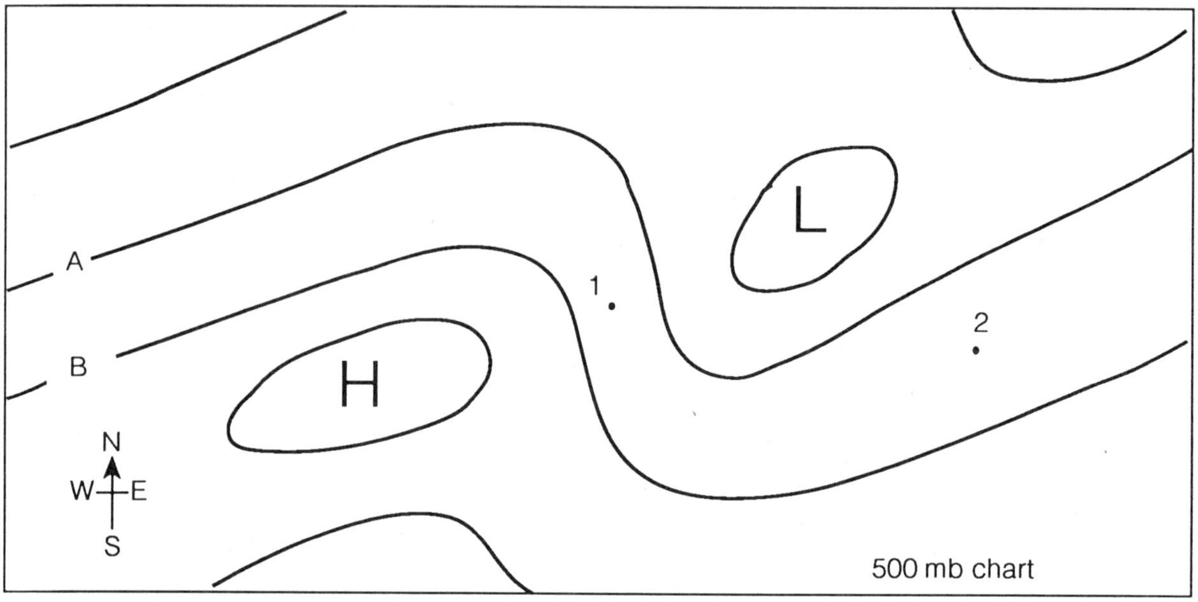

Figure 5

5. Answer the following questions that pertain to Figure 5, a 500 millibar chart.

a. The solid lines on the 500 millibar chart in Figure 5 are contour lines, indicating elevation above sea level. Which contour line (A or B) represents a higher altitude? __________

b. Which altitude below comes closest to the actual height above sea level of point 1 in Figure 5? (Hint: see Figure 10.3, p. 244 and Table 10.1, p. 246 in your textbook.)

1. 1,440 m, 4,800 ft
2. 3,000 m, 9,800 ft
3. 5,600 m, 18,000 ft
4. 9,180 m, 30,100 ft

c. Between contour line A and contour line B, draw arrows to show the most likely wind-flow pattern on the map.

d. Is the wind direction at point 1 most likely northwest, northeast, southwest, or southeast? __________ Is the wind direction at point 2 most likely northwest, northeast, southwest, or southeast? __________

e. Would the strongest winds on the map probably be occurring at point 1, point 2, the center of the L, or the center of the H? __________

f. Would the lowest height on the map be observed at point 1, point 2, the center of the L, or the center of the H? __________

g. Would the warmest air on the map most likely be observed at point 1, point 2, the center of the L, or the center of the H? __________

ADDITIONAL READINGS

"Simple Motions on the Rotating Earth" by Alfred Blackadar, *Weatherwise*, Vol. 39, No. 2 (April 1986), p. 99.

The Riddle of the Winds by W. S. Kales; Doubleday, Garden City, N.Y., 1977.

"Computer Map Analysis: Drawing Contours" by Alfred Blackadar, *Weatherwise*, Vol. 42, No. 2 (April 1989), p. 109.

The Restless Atmosphere by F. K. Hare; Hutchinson University Library, London, 1968.

ANSWERS

MATCHING

1. f
2. j
3. c
4. h
5. b
6. i
7. d
8. a
9. g
10. e

FILL IN THE BLANK

1. Coriolis force
2. isobars
3. Isaac Newton
4. pressure gradient force (PGF)
5. contour
6. left

MULTIPLE CHOICE

1. b
2. e
3. b
4. c
5. d
6. b
7. a
8. a
9. d

TRUE - FALSE

1. T
2. F
3. F
4. T
5. F
6. T
7. T
8. F
9. T
10. T
11. T
12. F
13. T
14. F
15. T

PROBLEMS

1. 1020 mb
2. a. low pressure area, N.H.
   b. high pressure area, S.H.
   c. high pressure area, N.H.
   d. low pressure area, S.H.
3. Around the low, the wind blows counterclockwise and parallel to the isobars or contours. Around the high, the wind blows clockwise and parallel to the isobars or contours.
4. a. 1020 mb
   b. northwest
   c. the L
   d. southeast
   e. point P; The isobars are closer together there, producing a stronger pressure gradient force.
   f. above the L, rising; above the H, sinking
5. a. contour B represents a higher altitude
   b. 3; 5,600 m or 18,400 ft
   c. the arrows should move from west to east parallel to the contour lines
   d. at point 1, northwest; at point 2, southwest
   e. point 1
   f. center of the L
   g. center of the H

CHAPTER

11

WIND: SMALL SCALE AND LOCAL SYSTEMS

Chapter Eleven focuses mainly on microscale (small scale) and mesoscale (middle scale) winds. The chapter begins by introducing the hierarchy of atmospheric motion, what scientists call the scales of motion. This is followed by a section that deals with friction and turbulence. The next several sections examine how the wind influences our environment. Here we see how the wind, blowing over a variety of surfaces, reshapes the landscape. After a section on wind instruments, we examine winds on a slightly larger scale. At this point we learn the causes of such winds as the sea breeze, the chinook, and the Santa Ana. We also learn that local winds that blow uphill during the day are called valley breezes, while those that blow downhill at night are called mountain breezes. At the end of the chapter is a section on desert winds and other local winds. Included here are the Texas norther and the northeaster.

Some important concepts of this chapter:

1. The friction layer is usually given as the first 1000 meters (3300 ft) above the surface.
2. A sudden change in wind speed or wind direction (or both) is called wind shear.
3. Winds that blow onshore, blow from the water onto the land; winds that blow offshore, blow from the land out over the water.
4. The prevailing wind is the direction from which the wind blows most frequently during a given time period.
5. The sea breeze (lake breeze) blows from the surface water onto the land in response to local pressure differences created by the uneven heating and cooling rates of land and water.
6. Monsoon winds are those that change direction seasonally. They usually blow from water onto land during the warm (wet) season, and from the land out over the water during the cool (dry) season.
7. Valley breezes blow uphill from the valley, while mountain breezes blow downhill from the mountain.
8. Chinook winds are warm, dry winds that warm by compressional heating as they descend the leeward side of mountains.
9. The Santa Ana wind is a warm, dry wind that warms by compressional heating as it descends the high plateau into Southern California.
10. Dust devils (whirlwinds) are rotating winds that usually form in clear, hot weather in dry areas. They are not tornadoes.

SELF TESTS

MATCH THE FOLLOWING

______ 1. Sudden change in wind speed or direction

______ 2. Indicates the percent of time the wind blows from different directions

______ 3. Protects crops and soil from the wind

______ 4. Eddies that form beneath the crest of a leewave

______ 5. Abbreviation for clear air turbulence

______ 6. Instrument that usually consists of three or more cups

______ 7. The friction of fluid flow

______ 8. Windblown clumps of snow often shaped like a cylinder

______ 9. A downdraft of an eddy that is usually observed in a region of strong vertical wind shear

______ 10. Instrument that indicates both wind speed and wind direction

______ 11. Another term for eddies created by obstructions

______ 12. Uses infrared or visible light in the form of a laser beam to determine wind information

______ 13. Wind blowing past a chimney is an example of this scale of motion

______ 14. Employs Doppler radar to obtain a vertical profile of wind speed and wind direction

______ 15. Water waves that oscillate back and forth in an open body of water

a. wind rose

b. snow rollers

c. rotors

d. wind profiler

e. aerovane

f. wind shear

g. lidar

h. viscosity

i. shelter belts

j. seiches

k. cup anemometer

l. mechanical turbulence

m. CAT

n. air pocket

o. microscale

■ ADDITIONAL MATCHING (WINDS)

_____ 1. Wind system that changes direction seasonally

_____ 2. Another name for a whirlwind

_____ 3. A chinook wind in the Alps

_____ 4. A strong, often cold, downslope wind

_____ 5. The nighttime counterpart of the sea breeze

_____ 6. Cumulus clouds developing above isolated mountain peaks are often the result of this

_____ 7. Warm, dry wind associated with devastating fires in Southern California

_____ 8. A windstorm composed of dust and sand that forms along the leading edge of a thunderstorm

_____ 9. Wind that brings relief from the bitter cold on the eastern side of the Rockies

_____ 10. Another name for the blue norther

_____ 11. Strong onshore coastal winds along the eastern seaboard of North America, often accompanied by heavy rain or snow

_____ 12. The leading edge of a sea breeze

a. foehn

b. haboob

c. northeaster

d. dust devil

e. Texas norther

f. land breeze

g. monsoon

h. Santa Ana

i. valley breeze

j. sea breeze front

k. chinook

l. katabatic wind

FILL IN THE BLANK

1. A cool, summertime breeze that blows from sea to land: __________ __________.

2. Name of the warm, dry downslope wind observed on the eastern side of the Rocky Mountains: __________ __________.

3. The smallest scale of motion is called the ______________.

4. Downslope winds warm by the process of ______________ heating.

5. The name given to the warm, dry downslope wind that blows from the east or northeast into Southern California: __________ __________ __________.

MULTIPLE CHOICE

1. In humid climates, clouds at night tend to form over the water during a:
   a. Santa Ana wind
   b. chinook wind
   c. lake breeze
   d. land breeze
   e. sea breeze

2. The winter monsoon in eastern and southern Asia is characterized by:
   a. wet weather and winds blowing from sea to land
   b. wet weather and winds blowing from land to sea
   c. dry weather and winds blowing from sea to land
   d. dry weather and winds blowing from land to sea

3. A thermal low does *not*:
   a. form in a region of warm surface air
   b. form in response to variations in surface air temperature
   c. become stronger with increasing height

4. Which of the following is an example of mesoscale motion?
   a. winds blowing past a row of trees
   b. winds blowing through a city
   c. winds on a 500-mb chart
   d. average winds of the Northern Hemisphere
   e. average winds of the entire world

5. The prevailing winds in South Florida are northeasterly. Consequently, the strongest sea breeze normally occurs on Florida's __________ coast, and the strongest land breeze on Florida's __________ coast.

   a. east, west
   b. east, east
   c. west, east
   d. west, west

6. An offshore wind blows:

   a. from water to land
   b. from land to water
   c. either from water to land or from land to water

7. During what time of day are surface winds normally strongest and most gusty?

   a. at sunset
   b. at sunrise
   c. several hours after sunrise
   d. several hours after sunset
   e. in the afternoon

8. Which of the winds listed below is *not* generally cold?

   a. California norther
   b. Texas norther
   c. bora
   d. mistral
   e. Columbia Gorge wind

9. Northeasters along the east coast of North America are best developed when a:

   a. high pressure area builds over the area
   b. low pressure area moves east over the Great Lakes
   c. low pressure area moves northeastward along the coast
   d. high pressure area moves south of the area

10. Which of the winds listed below is blowing uphill?

    a. chinook wind
    b. mountain breeze
    c. katabatic wind
    d. valley breeze
    e. Santa Ana wind

11. Which of the following is usually *not* observed during the passage of a summertime sea breeze?

    a. drop in air temperature
    b. drop in relative humidity
    c. wind shift

12. Clouds and precipitation are more likely to form on the downwind side of a large lake because on the downwind side the surface air tends to:

    a. diverge and rise
    b. diverge and sink
    c. converge and rise
    d. converge and sink

13. Suppose a west wind of 10 knots blows over a coastal region of densely covered shrubs. If this same wind moves out over the middle of a calm lake, its speed and direction would probably be:

    a. less than 10 knots and more southwesterly
    b. less than 10 knots and more northwesterly
    c. greater than 10 knots and more southwesterly
    d. greater than 10 knots and more northwesterly

14. Generally, the top of the friction layer is given to be about how high above the surface?

    a. 500 ft
    b. 1,000 ft
    c. 3,000 ft
    d. 10,000 ft
    e. 30,000 ft

■ TRUE - FALSE

_____ 1. Along the coast in summer, a sea breeze is usually strongest and best developed in the afternoon.

_____ 2. During a south wind, a wind vane will point toward the north.

_____ 3. A smog front is often associated with a sea breeze.

_____ 4. The speed of the wind generally increases with height above the earth's surface because only the lowest layer of air rotates with the earth.

_____ 5. The summer monsoon in eastern and southern Asia is characterized by wet weather and winds that blow from land to sea.

______ 6. A sea breeze is an example of macroscale winds.

______ 7. Clear air turbulence can occur in the vicinity of a jet stream.

______ 8. Dust devils usually descend from the base of a cumulonimbus cloud.

______ 9. Thermal turbulence above the surface is usually most severe at the time of maximum surface heating.

______ 10. Chinook wall clouds often signal the possible onset of a chinook.

______ 11. The so-called Boulder winds are gentle downslope winds that move through the town and surrounding area of Boulder, CO.

______ 12. On a clear, windy day the depth to which mixing occurs above the surface will depend upon the amount of surface heating.

■ PROBLEMS

1. Express the following wind directions in terms of compass points. (Hint: see Fig. 11.13, p. 279 in your textbook.)

Wind direction (degrees)	Wind direction (compass points)
270°	______________
360°	______________
90°	______________
225°	______________
315°	______________
135°	______________

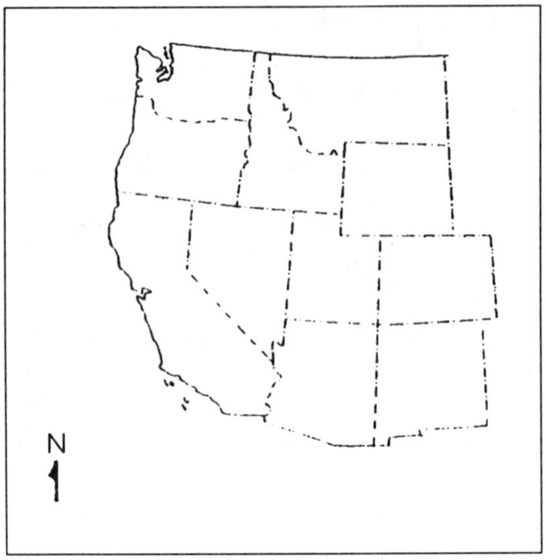

Figure 1

2. a. On the map (Figure 1), place an H where a high pressure area might be located during a strong Santa Ana wind.

   b. Draw several isobars around the high. Then draw several arrows to show how the winds would blow in Southern California.

   c. Even though the surface air in the center of the high may be cool, explain why the winds blowing into Southern California are warm and sometimes, downright hot.

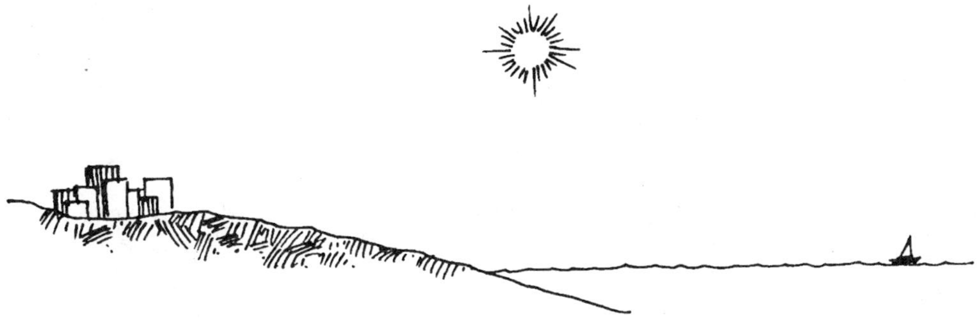

Figure 2

3. Figure 2 represents summertime conditions along the coast of a large body of water.

   a. In the diagram, label where the surface air is relatively warm and where it is relatively cool.

   b. Place an H in the diagram where the surface air pressure would be relatively high and an L where the surface air pressure would be relatively low.

   c. Draw several arrows to show the direction of the surface wind.

   d. The surface wind you have drawn in (d) would be called a __________ __________.

   e. Draw several more arrows and complete the thermal circulation.

   f. Draw a cloud to show where you would expect one to form. Explain why you chose that location.

■ ADDITIONAL READINGS

"Tornado or Dust Devil: The Enigma of Desert Whirlwinds" by Sherwood B. Idso, *American Scientist,* Vol. 62, No. 5 (Sept-Oct 1974), p. 530.

"Monsoons" by Priit J. Vesilind, *National Geographic,* Vol. 166, No. (December 1984), p. 712.

"The Santa Ana Wind of Southern California" by Arthur G . Lessard, *Weatherwise,* Vol. 41, No. 2 (April 1988), p. 100.

"The Blue Norther" by H. Michael Mogil, *Weatherwise,* Vol. 38, No. 3 (June 1985), p. 149.

"Utah's Great Salt Lake — A Classic Lake Effect Snowstorm" by David M. Carpenter, *Weatherwise,* Vol. 38, No. 6 (December 1985), p. 309.

"How Strong is the Wind" by Frank H. Forrester, *Weatherwise,* Vol. 39, No. 3 (June 1986), p. 147.

"Dust Devils in the Desert" by Andy Woodcock, *Weatherwise,* Vol. 44, No. 4 (Aug-Sept 1991), p. 39.

"Chinook Winds Resemble Water Flowing over A Rock" by Richard A. Kerr, *Science,* Vol. 231, No. 4743 (March 14, 1986), p. 1244.

"Front Range Windstorms Revisited: Small Scale Differences Amid Large Scale Similarities" by Edward J. Zipser and Alfred J. Bedard, Jr., *Weatherwise,* Vol. 35, No. 2 (April 1982), p. 82.

"Yellowstone Steam Devils" by Ronald L. Holle, *Weatherwise,* Vol. 35, No. 3 (June 1982), p. 115.

"Snow Rollers" by Francis M. Tam, *Weatherwise,* Vol. 35, No. 6 (December 1982), p. 276.

"The Puget Sound Convergence Zone" by Clifford Mass, *Weatherwise,* Vol. 33, No. 6 (December 1980), p. 272.

"Satellite Analyses of Antarctic Katabatic Wind Behavior" by David H. Bromwich, *Bulletin of the American Meteorological Society,* Vol. 70, No. 7 (July 1989), p. 738.

ANSWERS

MATCHING

1. f
2. a
3. i
4. c
5. m
6. k
7. h
8. b
9. n
10. e
11. l
12. g
13. o
14. d
15. j

ADDITIONAL MATCHING (WINDS)

1. g
2. d
3. a
4. l
5. f
6. i
7. h
8. b
9. k
10. e
11. c
12. j

FILL IN THE BLANK

1. sea breeze
2. chinook wind
3. microscale
4. compressional
5. Santa Ana wind

MULTIPLE CHOICE

1. d
2. d
3. c
4. b
5. a
6. b
7. e
8. a
9. c
10. d
11. b
12. c
13. d
14. c

TRUE - FALSE

1. T
2. F
3. T
4. F
5. F
6. F
7. T
8. F
9. T
10. T
11. F
12. T

PROBLEMS

1. 270° — west wind
   360° — north wind
   90° — east wind
   225° — southwest wind
   315° — northwest wind
   135° — southeast wind
2. a. On the map, the H should be located to the northeast of Southern CA.
   b. The winds should be blowing from a general east or northeast direction.
   c. The winds are warm because the air is moving downhill from a high plateau. Compressional heating warms the sinking air.

3. a. The relatively warm air is above the land; the relatively cool air is above the water.
   b. The surface H should appear above the water, the surface L above the land.
   c. The surface wind will blow from the water onto the land.
   d. Sea breeze (an onshore wind)
   e. To complete the circulation the air should be rising over the land. Aloft, the air should be moving from land to water, with air sinking over the water.
   f. The cloud should be drawn over the land where the air is rising, expanding and cooling.

CHAPTER 12

WINDS: GLOBAL SYSTEMS

Chapter Twelve examines the large-scale circulation of air. The chapter begins by looking at the global pattern of winds and pressure systems at the surface and aloft. Here we learn that there are semipermanent pressure systems that tend to persist in particular regions throughout the year. Winds that blow in response to these surface pressure systems establish prevailing wind patterns, such as the middle latitude westerlies, the polar easterlies, and the subtropical trade winds. The annual shifting of the pressure systems strongly influences the prevailing winds and annual precipitation patterns of many regions. Aloft we find westerly winds at middle and high latitudes in both hemispheres. After a section that describes the motion and formation of jet streams, the chapter details the many interactions that take place between the atmosphere and the oceans. Here we see that the interaction is an ongoing process where everything in one way or another seems to influence everything else.

Some important concepts of this chapter:

1. The two major semipermanent subtropical highs that influence the weather of North America are the Pacific high situated off the west coast and the Bermuda-Azores high situated off the southeast coast.
2. The subtropical high pressure areas with their sinking air and clear skies are primarily responsible for the major deserts of the world located near 30° latitude.
3. The polar front is a zone of low pressure where storms often form. It separates the mild westerlies from the cold, polar easterlies.
4. The trade winds are located equatorward of the subtropical highs in both hemispheres.
5. Near the equator, the intertropical convergence zone (ITCZ) is a boundary where air rises in response to the flowing together of the northeast trades and the southeast trades.
6. In the Northern Hemisphere, the major global pressure systems and wind belts shift northward in summer and southward in winter.
7. Warm air aloft (high pressure) over the tropics and subtropics, and cold air aloft (low pressure) over the middle and high latitudes produce westerly winds aloft in both hemispheres, especially in the middle and high latitudes.

8. Jet streams exist where strong winds become concentrated in narrow bands. They often form where sharp temperature changes produce rapid changes in pressure, such as aloft in the vicinity of the polar front.
9. Jet streams meander in a wavy west-to-east pattern, becoming strongest in winter when the contrast in temperature between high and low latitudes is greatest.
10. Surface winds blowing over the ocean drive the major ocean currents. The currents, in turn, release energy to the atmosphere which helps the atmosphere maintain its general circulation of winds.
11. Upwelling occurs along coastal areas where the prevailing winds blow parallel to the coast and the Coriolis force is able to bend the moving water seaward. Upwelling is especially prevalent along the west coast of North and South America.
12. A major El Niño event is a condition where warm surface water covers vast areas of the tropical Pacific, especially off the coast of South America.

SELF TESTS

MATCH THE FOLLOWING

______	1.	Fast flowing current of air concentrated in a narrow band	a.	polar easterlies
______	2.	Wind belt observed behind the polar front	b.	oceanic front
______	3.	Pressure systems that almost circle the globe near latitude 30°	c.	ITCZ
______	4.	Ocean current that flows parallel to the west coast of North America	d.	California current
______	5.	Lines of equal wind speed	e.	upwelling
______	6.	The shallow, cold surface anticyclone observed over Asia during winter	f.	subpolar low
______	7.	Semipermanent pressure system associated with the polar front	g.	Labrador current
______	8.	The turning of water with depth	h.	jet stream
______	9.	Boundary separating two contrasting water masses	i.	Gulf Stream
______	10.	The reversal of surface air pressure at opposite ends of the tropical Pacific ocean	j.	Southern Oscillation
______	11.	Ocean current that brings cold water southward along the east coast of North America	k.	Siberian high
______	12.	The rising of cold water	l.	Ekman spiral
______	13.	Warming of the surface water of the tropical Pacific west of South America	m.	subtropical highs
______	14.	Warm ocean current that flows northward along the east coast of North America	n.	El Niño
______	15.	Boundary separating the northeast trades and the southeast trades	o.	isotachs

FILL IN THE BLANK

1. The majority of the United States lies within a wind belt known as the _________.

2. _________ is mainly responsible for the cold coastal water observed along the west coast of North America.

3. The jet stream situated at the tropopause in the vicinity of the polar front is called the _________ _________ _________.

4. The Pacific Ocean's counterpart to the Atlantic's Bermuda-Azores high is called the _________ _________.

5. Much of Alaska lies within this wind belt: _________ _________.

6. The cold-water episode that finds strong trade winds and unusually cold surface water in the central and eastern Pacific is called _________.

7. _________ _________ are responsible for the existence of the major deserts of the world observed near 30° latitude.

MULTIPLE CHOICE

1. The horse latitudes are the result of:

   a. the polar front jet stream
   b. the ITCZ
   c. the subtropical highs
   d. the subpolar lows
   e. the polar high

2. In terms of the three-cell model of the general circulation, areas of surface low pressure should be found near:

   a. the equator and 60° latitude
   b. the equator and the poles
   c. the equator and 30° latitude
   d. 30° latitude and 60° latitude
   e. 30° latitude and the poles

3. The main reason for the dry summers observed along the southwest coast of North America is that:

   a. the air along the coast is too cool to produce adequate precipitation
   b. the ITCZ moves north in summer, blocking storms from entering the area
   c. the polar jet stream moves over the area in summer
   d. the northeast trades sweep into the region
   e. the Pacific high moves northward, producing sinking air over the region

4. A jet streak is:

   a. a contrail that forms behind a jet aircraft
   b. a region of strong winds in a jet stream
   c. a sharp change in wind speed or wind direction
   d. the flow of air around a mountain
   e. a small jet stream less than 1000 km long

5. At the equator, according to the three-cell general circulation model, we would *not* expect to find:

   a. heavy showers
   b. the ITCZ
   c. cumuliform clouds
   d. a ridge of high pressure

6. Which below is *not* considered a semipermanent high or low pressure area?

   a. Siberian high
   b. Aleutian low
   c. Icelandic low
   d. Bermuda high
   e. Pacific high

7. The three-cell model of the general circulation says that in the Northern Hemisphere, you would expect to observe the westerlies:

   a. southward of the ITCZ
   b. southward of the subtropical highs
   c. northward of the subtropical highs
   d. northward of the subpolar lows
   e. southward of the northeast trades

8. According to the three-cell general circulation model, one would expect the driest regions of the world to be near:

   a. latitude 30° and latitude 60°
   b. latitude 30° and the polar regions
   c. the equator and the polar regions
   d. the equator and 30° latitude
   e. the equator and 60° latitude

9. Which of the following does *not* occur during a major El Niño event?

   a. extensive ocean warming occurs over the tropical Pacific Ocean
   b. it may last for many months
   c. large numbers of fish often die along the coast of Peru
   d. it influences the westerly winds aloft, bringing too much rain to some regions and too little to others
   e. the subtropical high pressure area off the coast of South America increases in strength

10. An air parcel moving northward from the equator moves closer to the earth's axis of rotation. Because of the conservation of angular momentum, the parcel's speed should:

    a. increase
    b. decrease
    c. show no change

11. In the middle and high latitudes, the winds aloft tend to blow:

    a. from north to south
    b. from south to north
    c. from east to west
    d. from west to east

TRUE - FALSE

______ 1. The trade winds are found equatorward of the westerlies in both hemispheres.

______ 2. The polar jet and the subtropical jet are both examples of tropopause jets.

______ 3. In the Northern Hemisphere, the major features of the general circulation shift northward in summer and southward in winter.

______ 4. The polar front jet stream is strongest in winter but moves farther south in summer.

______ 5. In the three-cell model of the general circulation, the Hadley cell is a thermal circulation driven by convective "hot" towers along the equator.

______ 6. Normally, winds aloft are strongest during the winter.

______ 7. The polar front jet is normally observed at a higher altitude than the subtropical jet.

______ 8. Average winter temperatures in Great Britain and Norway would probably be much colder if it were not for the Canary current.

______ 9. The polar front separates the cold, polar easterlies from the milder westerlies.

______ 10. The polar jet stream tends to flow in a wavy east-to-west pattern.

______ 11. The low-level jet stream that forms over the Central Plains of the United States appears to be responsible for the formation of nighttime thunderstorms.

______ 12. Air moving eastward more slowly than the earth rotates would be an east wind to an observer on the earth.

______ 13. The polar front jet is normally observed equatorward of the subtropical jet.

______ 14. In the Northern Hemisphere, the polar front jet stream actually separates colder air to the north from relatively mild air to the south.

______ 15. The winds aloft are strongest over the middle latitudes of the Northern Hemisphere when, aloft, very cold air is to the north and very warm air is to the south.

PROBLEMS

1. In the circle representing the earth (Figure 1), place the major semi-permanent surface pressure systems at their appropriate latitudes. Label them properly and be sure to include the polar front and the ITCZ. (Hint: Use the three-cell model of the general circulation.)

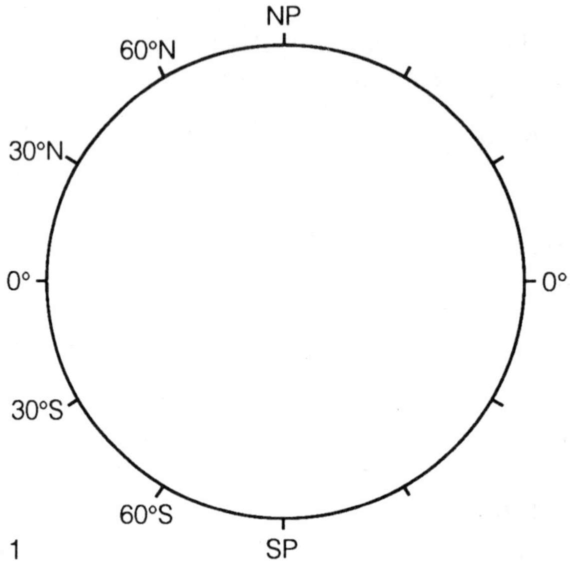

Figure 1

2. In the circle representing the earth (Figure 2) draw arrows to represent the major surface wind belts of the world. Label them correctly. (Hint: Again use the three-cell model.)

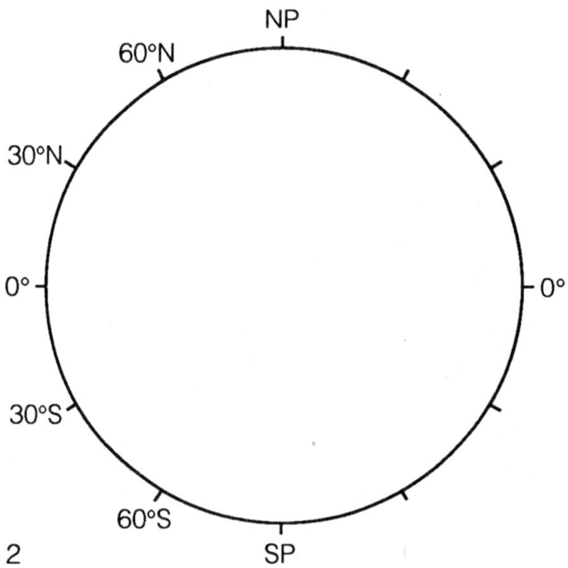

Figure 2

3. Use Figures 1 and 2 and the three-cell model of the general circulation of the atmosphere to predict the prevailing winds for each of the following cities:

	Prevailing Winds
a. Chicago, Illinois (latitude 42°N)	______________
b. Barrow, Alaska (latitude 70°N)	______________
c. Winnipeg, Canada (latitude 50°N)	______________
d. Miami, Florida (latitude 26°N)	______________
e. Auckland, New Zealand (latitude $37^1/_2$°S)	______________
f. Launda, Angola (latitude 9°S)	______________

■ ADDITIONAL READINGS

"Angular Momentum Cycle in the Atmosphere-Ocean-Solid Earth System" by Abraham H. Oort, *Bulletin of the American Meteorological Society,* Vol . 70, No. 10 (October 1989), p. 1231.

"El Niño and La Niña" by George Philander, *American Scientist,* Vol. 77, No. 5 (Sept-Oct 1989), p. 451.

"Written in the Winds" by Jerome Namias, *Weatherwise,* Vol. 42, No. 2 (April 1989), p. 85.

"Ocean Color — A Key to Climatic Change" by Patrick Hughes, *Weatherwise,* Vol. 39, No. 5 (October 1986), p. 267.

"A History of the Prevailing Ideas About the General Circulation of the Atmosphere" by Edward N. Lorenz, *Bulletin of the American Meteorological Society,* Vol. 64, No. 7 (July 1983), p. 730.

"El Niño: The Great Equatorial Pacific Ocean Warming Event of 1982-83" by Eugene M. Rasmusson and J. Michael Hall, *Weatherwise,* Vol. 36, No. 4 (August 1983), p. 166.

"Motion in the Ocean" by George Alexander, *Weatherwise,* Vol. 33, No. 6 (December 1980), p. 265.

"All About the Jet Stream" by Thomas Schlatter, *Weatherwise,* Vol. 40, No. 1 (February 1987), p. 50.

"North Atlantic Oscillation: Concept and an Application" by Peter J. Lamb and Randy A. Peppler, *Bulletin of the American Meteoroloical Society,* Vol. 68, No. 10 (October 1987), p. 1218.

ANSWERS

MATCHING

1. h
2. a
3. m
4. d
5. o
6. k
7. f
8. l
9. b
10. j
11. g
12. e
13. n
14. i
15. c

FILL IN THE BLANK

1. westerlies
2. upwelling
3. polar front jet stream, or polar jet stream
4. Pacific high or Pacific anticyclone
5. polar easterlies
6. La Niña
7. subtropical highs or subtropical anticyclones

MULTIPLE CHOICE

1. c
2. a
3. e
4. b
5. d
6. a
7. c
8. b
9. e
10. a
11. d

TRUE - FALSE

1. T
2. T
3. T
4. F
5. T
6. T
7. F
8. F
9. T
10. F
11. T
12. T
13. F
14. T
15. T

PROBLEMS

1 and 2. Answers are found on p. 305, Figure 12.2, in your textbook.

3. Predicted prevailing winds
   a. westerly
   b. easterly or northeasterly
   c. westerly
   d. northeasterly
   e. westerly
   f. southeasterly

CHAPTER

13

AIR MASSES AND FRONTS

The first part of Chapter Thirteen examines the various air masses and the weather each brings to a particular region. Here we learn that an air mass has fairly uniform properties of temperature and moisture in any horizontal direction, and for an air mass to acquire such properties it must originate over generally flat terrain of uniform composition, where the winds are fairly light. Where air masses with sharply contrasting properties meet, we find weather fronts. Consequently, the emphasis of the latter part of the chapter is on the structure, weather, and development of fronts.

Some important concepts of this chapter:

1. An air mass is a large body of air with similar horizontal temperature and moisture characteristics.
2. Regions where air masses originate and acquire their properties of temperature and moisture are called source regions. These source regions are generally flat, of uniform composition, where winds are light.
3. Continental (c) air masses form over land. Maritime (m) air masses form over water. Polar (P) air masses originate in cold, polar latitudes, and extremely cold air masses are designated as arctic (A). Tropical (T) air masses originate in warm, tropical latitudes.
4. cP air masses are cold and dry; cA air masses are extremely cold and dry; cT air masses are hot and dry; mT air masses are warm and moist; mP air masses are cold and moist.
5. A front is a transition zone that separates two air masses with contrasting properties, usually temperature and humidity.
6. A cold front is a region where colder air is replacing warmer air. Typically, along the leading edge of the cold front, warmer air is forced upward producing a rather narrow band of showers.
7. The cold front represents a trough of lower pressure where there is a wind shift (often from SW to NW). After the passage of a cold front, air temperatures drop and air pressures, often having fallen, begin to rise.
8. A warm front is a region where warmer air is replacing colder air. Along the warm front, warmer air rides up and over colder surface air, producing clouds and precipitation out ahead of the advancing surface front.
9. As a warm front approaches, the clouds often change from high clouds (Ci and Cs), to middle clouds (As), to low clouds (Ns and St).
10. An occluded front often has characteristics of both a cold front and a warm front. The coldest air is observed *behind* a cold occlusion and *ahead* of a warm occlusion.

SELF TESTS

MATCH THE FOLLOWING

(Note: Some answers will be used more than once.)

_____ 1. Air mass responsible for hot, muggy weather in the eastern half of North America

_____ 2. The coldest of all air masses

_____ 3. This front is drawn in red on a weather map

_____ 4. The rising of warm air over cold air

_____ 5. A warm, moist air mass that forms over water

_____ 6. An air mass that forms over North America only in summer

_____ 7. Air mass responsible for refreshing cool, dry breezes after a long, humid summer hot spell in the Central Plains

_____ 8. Air mass responsible for heavy rain, flooding mudslides and the melting of snow at high elevations in Southern California

_____ 9. Suggests that a front is regenerating or strengthening

_____ 10. This front is drawn in purple on a weather map

_____ 11. Air mass responsible for hot, dry summer weather over portions of Arizona and New Mexico

_____ 12. During the winter this air mass is responsible for heavy snow showers on the western slopes of the Rockies

_____ 13. This front is drawn in blue on a weather map

_____ 14. Air mass responsible for daily afternoon thunderstorms along the Gulf Coast states

_____ 15. Record-breaking low temperatures during the middle of winter are most often associated with this air mass

_____ 16. Cold, moist air mass that moves into northeastern North America from off the Atlantic Ocean

a. warm front

b. cT

c. occluded front

d. mT

e. cold front

f. cP

g. frontogenesis

h. cA

i. mP

j. overrunning

■ FILL IN THE BLANK

1. In the space next to each air mass letter designation, write in what the letters represent. Also describe the temperature and moisture characteristics of the air mass.

Letter designation	Name	Temperature/Moisture Characteristics
cP		
cT		
cA		
mP		
mT		

2. The heavy snow showers that fall on the eastern shores of the Great Lakes are called __________ __________.

3. On a surface weather map, the transition zone between two air masses with sharply contrasting properties is called a __________.

4. An extremely large body of air whose properties of temperature and moisture are similar in any horizontal direction: __________ __________.

5. On a surface weather map a __________ front represents a region where colder air is replacing warmer air.

6. On a surface weather map where cold air is replacing cool air, an __________ front is drawn.

7. On a surface weather map a __________ front represents a region where warmer air is replacing colder air.

MULTIPLE CHOICE

1. As you travel toward a warm front during the winter, the *most likely* sequence of weather you would experience is:

   a. snow, freezing rain, hail, rain
   b. snow, sleet, freezing rain, rain
   c. snow, freezing rain, hail, sleet
   d. rain, snow, sleet, freezing rain
   e. freezing rain, snow, sleet, rain

2. The origin of cA and cP air masses that enter the United States is:

   a. the North Atlantic Ocean
   b. northern Siberia
   c. northern Canada and Alaska
   d. the North Pacific Ocean
   e. the desert southwest

3. During the winter, cold, dry air will occasionally move into Washington, Oregon, and California from the east and northeast. However, by the time this cold air reaches the coastal regions it is often much warmer than it was originally primarily because:

   a. the air sinks, compresses and warms
   b. friction with the ground warms the air
   c. the sun heats the air
   d. the ocean warms the air
   e. latent heat of condensation warms the air as it moves downhill

4. The greatest contrast in both *temperature* and *moisture* will occur along the boundary separating which air masses?

   a. cP and cT in summer
   b. cP and mT in summer
   c. mP and mT in winter
   d. mP and mT in winter
   e. cA and mT in winter

5. As an air mass moves over a large warm lake it will undergo the most drastic change in both *temperature* and *humidity* when it is originally classified as:

   a. cA
   b. cT
   c. cP
   d. mT
   e. mP

6. Lake-effect snows are:

   a. heavy snows that accumulate on the upwind side of a lake
   b. snowstorms that form along the leading edge of a cold front in the vicinity of the Great Lakes
   c. localized snowstorms that form on the downwind side of a lake
   d. warm frontal snows that are produced by cool mP air flowing over a lake
   e. surprise spring snows that most often occur during March and April, after a large lake is frozen.

7. What type of weather front would be responsible for the following weather forecast: "Increasing high cloudiness and cold this morning. Clouds increasing and lowering this afternoon with a chance of snow or rain tonight. Precipitation ending by noon tomorrow. Turning much warmer. Winds light easterly today becoming southeasterly tonight and southwesterly by tomorrow."

   a. warm front
   b. cold front
   c. warm-type occluded front
   d. cold-type occluded front

8. What type of weather front would be responsible for the following weather forecast: "Increasing cloudiness and warm today with the possibility of showers and thunderstorms by this evening. Turning much colder tonight. Winds southwesterly today, becoming gusty and shifting to northwesterly by tonight."

   a. cold-type occluded front
   b. warm-type occluded front
   c. warm front
   d. cold front

9. Typically, winter mP air masses along the Atlantic coast of North America are less common than mP air masses along the Pacific coast mainly due to the fact that:

   a. the water is warmer along the Atlantic coast
   b. the prevailing winds aloft are westerly
   c. the landmass along the Atlantic coast is colder
   d. the source region for mP air on the Atlantic coast is western Europe
   e. the water is much colder along the Pacific coast

10. During the winter the upper air flow on the adjacent map would bring _________ air masses into western North America and _________ air masses into eastern North America.

    a. mP, mT
    b. cA, mP
    c. cA, mT
    d. cP, mP
    e. cA, cT

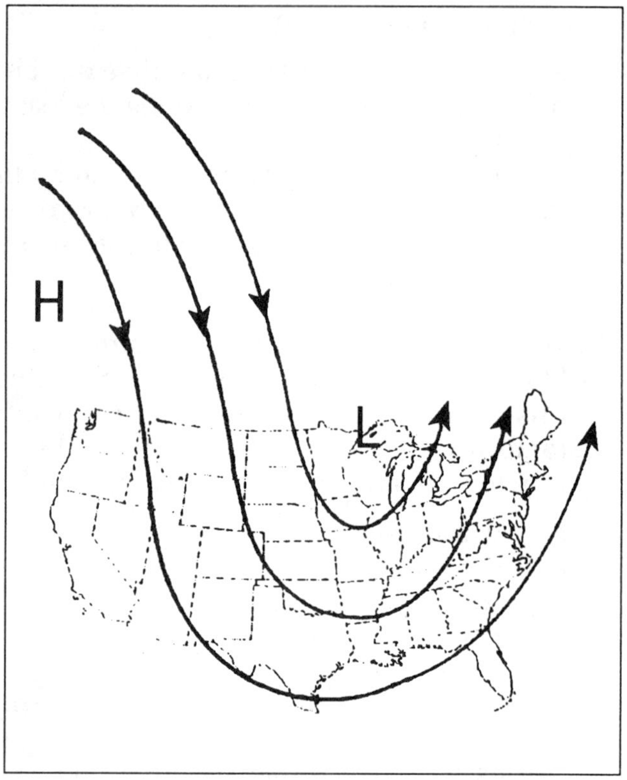

■ TRUE - FALSE

______ 1. On a surface weather map, a cold front represents a trough of low pressure.

______ 2. Generally, warm fronts move faster than cold fronts.

______ 3. A Texas norther (or blue norther) is most often associated with a warm front.

______ 4. Wintertime thunderstorms are most often associated with cold fronts.

______ 5. A good source region for an air mass is generally a windy region.

______ 6. Before the passage of a cold front the air pressure normally drops; after the passage, the air pressure normally rises.

______ 7. Occluded fronts may form as a cold front overtakes a warm front.

______ 8. On a surface weather map, a stationary front is drawn in purple.

______ 9. A moist, tropical air mass that is warmer than the surface over which it is moving would be classified as mTw.

______ 10. A back-door cold front moving through New England would most likely be moving from west to east.

______ 11. At the surface, during the passage of a warm-type occluded front, cooler air is being replaced by milder air.

______ 12. At the surface, during the passage of a cold-type occluded front, the coldest air is observed ahead of the advancing front.

______ 13. Persistent hot, humid weather that may last for days in the southeastern United States could be classified as air mass weather.

______ 14. Lake effect snows are most prevalent when the lakes are covered with ice.

______ 15. The slope of a typical warm front is usually much steeper than that of a cold front.

PROBLEMS

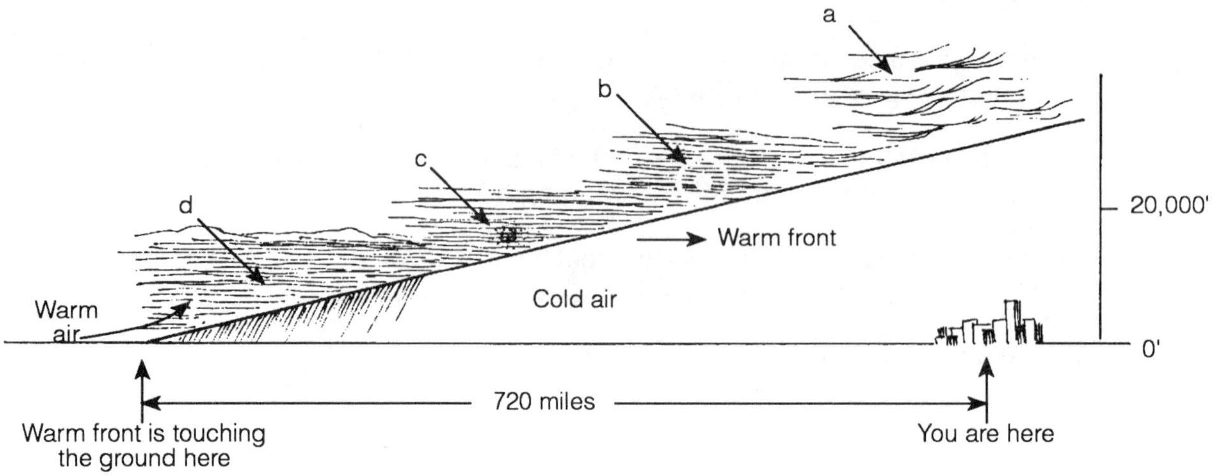

Figure 1

1. Figure 1 shows a warm front with precipitation moving toward your home in winter.

   a. Letters a through d represent a sequence of clouds that you might expect to observe as the warm front approaches you. List the clouds in their proper order below.

      cloud a ________________

      cloud b ________________

      cloud c ________________

      cloud d ________________

   b. Suppose that 720 miles from your home the warm front is at the surface. If the front is moving toward you at an average speed of 10 miles per hour, how long will it take before the front passes your area?

   c. If the band of precipitation extends out ahead of the surface front by about 240 miles, how long will it be before precipitation begins to fall at your home?

d. Suppose the leading edge of the warm front in Figure 1 is located about 24,000 ft (4.5 miles) above you. What would be the slope of this front?

Slope 1: __________

How would this slope compare to a "typical" warm front?

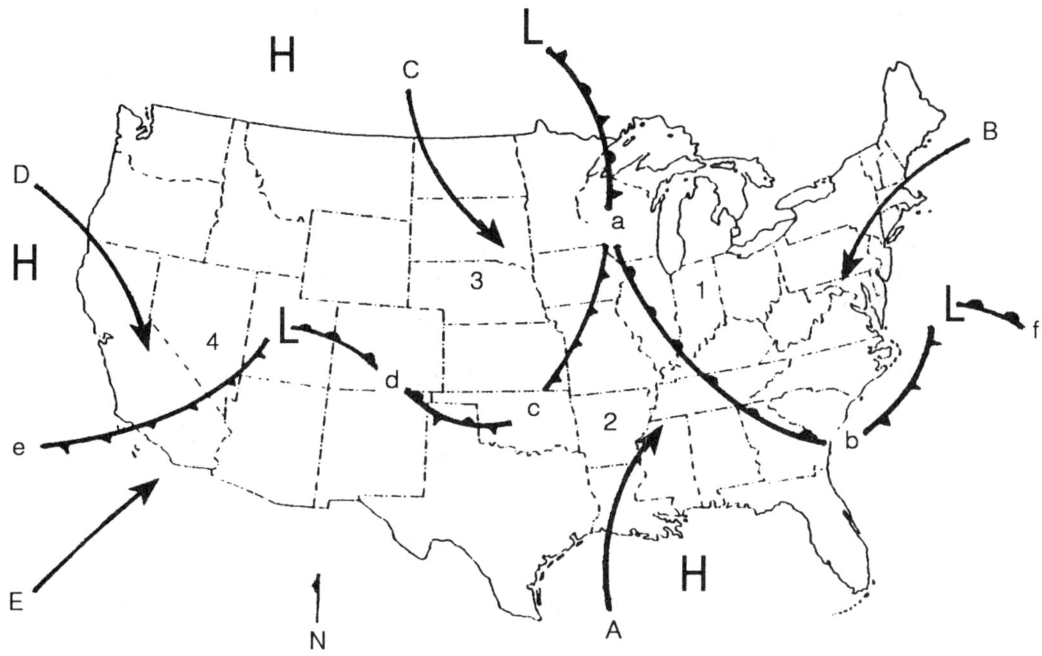

Figure 2

2. Figure 2 is a surface weather map. In the space below, label the fronts that fall between the letters on the map.

Front between letters	Name of front
L and a	____________
L and b	____________
L and f	____________
L and e	____________
L and d	____________
a and b	____________
a and c	____________
c and d	____________

3. Figure 2 represents a winter weather map with the large arrows showing the movement of air masses. The letters below correspond to the letters on the map. Next to each letter place the type of air mass that is present on the map and its temperature and moisture characteristic.

	Air Mass	**Characteristics**
A	_______	____________________
B	_______	____________________
C	_______	____________________
D	_______	____________________
E	_______	____________________

4. Look closely at Figure 2 and match the weather conditions listed below with the number that appears on the map.

   a. Overcast, cold, sleet, southeasterly winds, falling air pressure. Number _____

   b. Partly cloudy, windy, very cold, northwesterly winds, rising air pressure. Number _____

   c. Partly cloudy, mild, southwesterly winds, falling air pressure. Number _____

   d. Rainshowers, cool, northwesterly winds, rising air pressure. Number _____

5. In Chapter 6 (Humidity) we learned that the dew-point temperature is a good measure of the amount of water vapor in the air. With this information look at Figure 2 and determine:

   a. Which city (number) on the map would most likely have the *lowest* dew-point temperature? Number _____ Explain why.

   b. Which city (number) on the map would most likely have the *highest* dew-point temperature? Number _____ Explain why.

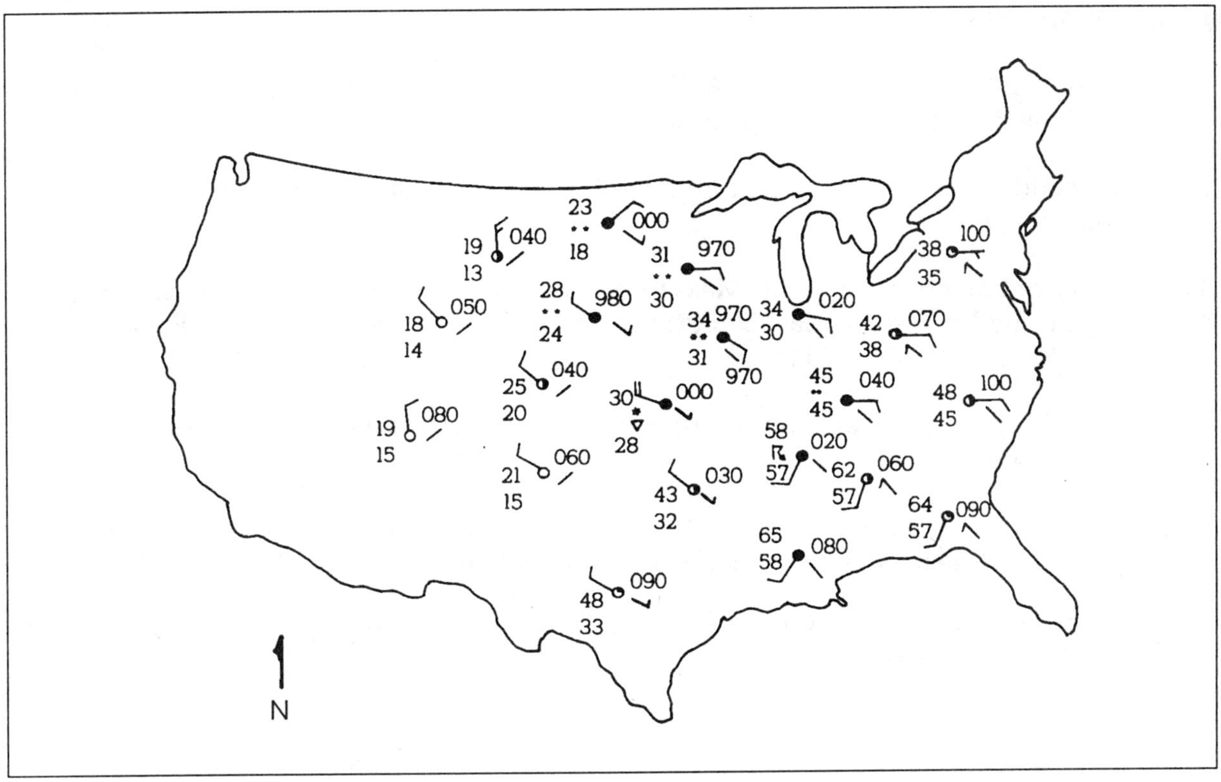

Figure 3

6. On the surface weather map above (Figure 3) locate and draw:

   a. the cold front
   b. the warm front
   c. the occluded front

   (Hint: the summary of information found in Table 13.2, p. 346; Table 13.3 p. 348; and Table 13.4, p. 350 should be most useful in helping you locate the fronts. Also, the weather for the various cities on the map — the station model information — is given in appendix B, p. 524, in your textbook.)

ADDITIONAL READINGS

"The Record Southeast Drought of 1986" by Kenneth H. Bergman, Chester F. Ropelewski and Michael S. Halpert, *Weatherwise,* Vol. 39, No. 5 (October 1986), p. 262.

"A Visible Cold Front" by Lynette Rummel, *Weatherwise*, Vol. 40, No. 4 (July-Aug 1987), p. 183.

"The History of Polar Front and Air Mass Concepts in the United States — An Eyewitness Account" by Jerome Namias, *Bulletin of the American Meteorological Society,* Vol. 64, No. 7 (July 1990), p. 734.

"Worst Heat Wave in 26 Years" by A. James Wagner, *Weatherwise*, Vol. 33, No. 4 (August 1980), p. 168.

"The Great Freeze of '83 — Analyzing the Causes" by H. Michael Mogil, Andrew Stern and Richard Hagan, *Weatherwise*, Vol. 37, No. 6 (December 1984), p.304.

"The Year the Weather Went Wild" by Thomas Y. Canby, *National Geographic*, Vol. 152, No. 6 (December 1977), p. 799.

ANSWERS

MATCHING

1. d
2. h
3. a
4. j
5. d
6. b
7. f
8. d
9. g
10. c
11. b
12. i
13. e
14. d
15. h
16. i

FILL IN THE BLANK

1. cP, continental polar, cold, dry
   cT, continental tropical, hot, dry
   cA, continental arctic, very cold, dry
   mP, maritime polar, cold, moist
   mT, maritime tropical, warm, moist
2. lake-effect snows
3. front
4. air mass
5. cold
6. occluded
7. warm

MULTIPLE CHOICE

1. b
2. c
3. a
4. e
5. a
6. c
7. a
8. d
9. b
10. c

TRUE - FALSE

1. T
2. F
3. F
4. T
5. F
6. T
7. T
8. F
9. T
10. F
11. T
12. F
13. T
14. F
15. F

PROBLEMS

1. a. cloud a, cirrus
   cloud b, cirrostratus
   cloud c, altostratus
   cloud d, nimbostratus
   b. it will pass in about 72 hours or 3 days
   c. precipitation should begin in about 48 hours or 2 days
   d. slope 1: 160, about average for a warm front
2. L and a, occluded front
   L and b, cold front
   L and f, warm front
   L and e, cold front
   L and d, warm front
   a and b, warm front
   a and c, cold front
   c and d, stationary front

3. A. maritime tropical, mT, warm and humid
   B. maritime polar, mP, cold and moist
   C. continental Arctic (or polar), cA or cP, cold and dry
   D. maritime polar, mP, cold and moist
   E. maritime tropical mT, warm and humid
4. a. number 1
   b. number 3
   c. number 2
   d. number 4
5. a. Lowest dew-point temperature at number 3. Cold, dry arctic (or polar) air from northern latitudes has moved into this region.
   b. Highest dew-point temperature at number 2. Warm, humid air from the Gulf of Mexico has moved into this region.
6. (See figure below.)

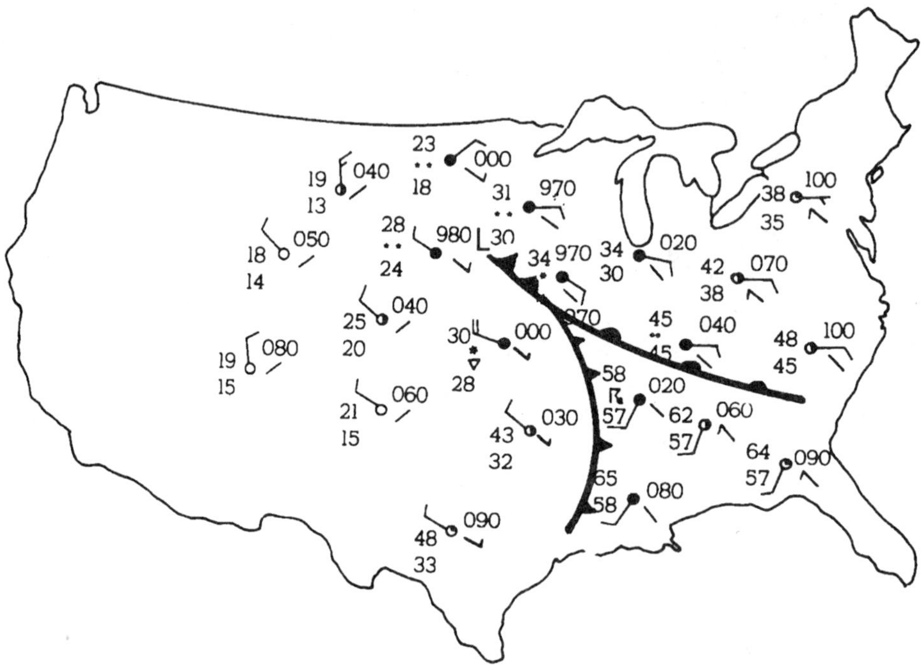

CHAPTER

14

MIDDLE LATITUDE STORMS

Chapter Fourteen describes where, why, and how a wave cyclone forms and moves along a front. The chapter begins by examining the early polar front theory proposed by Norwegian meteorologists after World War I. It then looks at the important effect the upper-air flow, including the jet stream, has on the intensification and movement of surface pressure systems. Here we learn that along the flanks of a jet stream, regions of convergence supply air to surface anticyclones, and regions of divergence remove air above surface midlatitude cyclones. The curving nature of the jet stream tends to direct anticyclones southeastward and surface lows northeastward. The chapter concludes by looking at the concept of vorticity and how it relates to developing middle latitude cyclones, longwaves, and lee-side lows.

Some important concepts of this chapter:

1. Conceived in Norway, the polar front theory is a model of how an ideal storm progresses through the stages of birth, maturity, and finally, dissipation.
2. For a surface midlatitude storm to develop or intensify, the upper-level low must be located to the west of (behind) the surface low.
3. Convergence is the piling up of air; divergence is the spreading out of air.
4. For a surface midlatitude storm to develop or intensify, upper-level divergence above the storm must be greater than surface convergence (more air must be removed at the top than is brought in at the surface).
5. For a surface anticyclone to build (increase in strength), upper-level convergence above the high must be greater than surface divergence (inflow of air at the top of the system must be greater than outflow of air at the surface).
6. As a jet stream develops into a looping wave it provides areas of convergence for the surface anticyclones and areas of divergence for the surface midlatitude cyclones.
7. The curving nature of a jet stream tends to direct anticyclones southeastward and midlatitude storms northeastward.
8. When the upper-air flow is disturbed by a shortwave, horizontal and vertical air motions begin to enhance the formation of a surface storm as the rising of warm air and the sinking of cold air provide the proper energy conversion for the storm's growth.

9. Vorticity is a measure of the spin (usually horizontal) of small air parcels. When an area of positive vorticity approaches an area, it is usually accompanied by diverging air aloft, converging air at the surface, and cyclonic storm development.

10. As air flows over a mountain range, changes in the air column cause storms to either intensify or form on the downwind (leeward) side of the mountain.

SELF TESTS

MATCHING

______ 1. The development or strengthening of a midlatitude storm system

______ 2. The piling up of air above a region

______ 3. The spin of air parcels

______ 4. A kink that forms along a front

______ 5. An upper-level cold pool of air that has broken away from the main flow

______ 6. A term used by meteorologist for explosive cyclogenesis

______ 7. Small disturbances imbedded in a longwave

______ 8. Another name for a large midlatitude cyclonic storm

______ 9. Another term for Rossby waves

______ 10. An atmospheric condition, where, on an upper-level chart, isotherms cross the isobars (or contours) and temperature advection occurs

______ 11. The spreading out of air above a region

______ 12. A trough that moves westward instead of eastward is said to possess this type of motion

a. convergence
b. shortwave
c. cyclogenesis
d. baroclinic
e. frontal wave
f. divergence
g. bomb
h. longwaves
i. retrograde
j. cut-off low
k. wave cyclone
l. vorticity

FILL IN THE BLANK

1. For a surface midlatitude storm to develop or intensify into a deep low pressure area, aloft __________ must be greater than the surface convergence of air.

2. __________ instability is needed for a surface low pressure area to intensify into a huge storm.

3. Storm systems that tend to form on the downwind side of a mountain range are called __________ __________.

4. When a surface anticyclone is building or strengthening, aloft __________ must be greater than surface divergence of air.

5. As the __________ __________ bends into a looping wave pattern, it provides some of the necessary ingredients for a developing surface storm system.

6. For a surface midlatitude storm to develop or intensify, the upper level low must be to the __________ of the surface low.

MULTIPLE CHOICE

1. When an upper-level low lies directly above a midlatitude storm system, the surface low will usually:

   a. dissipate
   b. intensify
   c. show no change during a 48 hour period

2. The polar front theory was developed:

   a. shortly after the end of World War II
   b. just before the turn of the 19th century
   c. shortly after the end of the Korean War
   d. shortly after the end of World War I

3. If an upper-level trough is located to the west of a surface midlatitude storm, the surface storm will probably move toward the:

   a. southwest
   b. northeast
   c. northwest

4. Which below is *not* a name given to a large cyclonic storm system that forms in the middle latitudes?

   a. wave cyclone
   b. middle latitude cyclone
   c. anticyclone
   d. extratropical cyclone

5. If we assume that the absolute vorticity of flowing air is conserved, air moving southeastward will bend __________ to compensate for the __________ in the earth's vorticity.

   a. cyclonically, decrease
   b. cyclonically, increase
   c. anticyclonically, decrease
   d. anticyclonically, increase

6. According to the polar front theory of a developing wave cyclone, the storm system is usually most intense:

   a. as a stable wave
   b. as a stationary front
   c. as an open wave
   d. as a frontal wave
   e. when the system first becomes occluded

■ TRUE - FALSE

______ 1. The rising of warm air and the sinking of cold air provide energy for a developing wave cyclone.

______ 2. When upper-level convergence of air above a large area of surface high pressure is greater than surface divergence of air, the pressure at the center of the anticyclone will rise.

______ 3. Storms generally move faster in summer than in winter.

______ 4. The earth's vorticity is zero at the poles.

______ 5. When a high pressure area (anticyclone) is said to be building, the air pressure at the center of the high is rising.

______ 6. The polar front theory was conceived in the United States.

______ 7. In the Northern Hemisphere, air that spins cyclonically possesses positive vorticity.

______ 8. When a mid-latitude storm is said to be deepening or intensifying, the air pressure at the center of the storm is dropping.

______ 9. The eastern slope of the Rockies is frequently a region of cyclogenesis.

______ 10. Aloft, shortwaves usually move slower than longwaves.

______ 11. Aloft, shortwaves usually weaken when they approach a ridge.

■ PROBLEMS

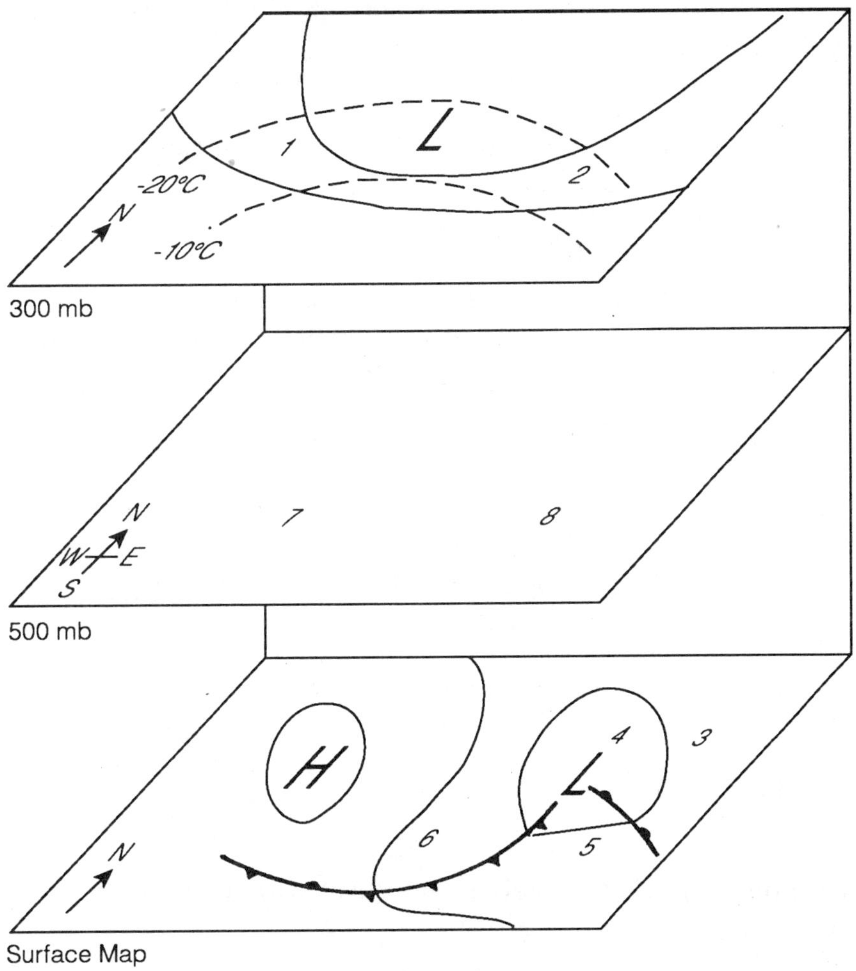

Figure 1

Figure 1 shows a surface map, a blank 500-mb chart, and a 300-mb chart. The solid lines on the surface map represent isobars. The solid lines on the 300-mb chart represent contours, while the dashed lines are isotherms.

1. In Figure 1 on the 300-mb chart:

   a. Draw several arrows that represent the wind-flow pattern at this level. From this, determine the wind direction at point 1 and at point 2.

      Wind direction at point 1 ________

      Wind direction at point 2 ________

   b. If the winds on the 300-mb chart are blowing at a constant speed parallel to the contour lines, then convergence of air is probably occurring at point ________, and divergence at point ________.

   c. From the wind-flow pattern on the 300-mb chart, it appears that the surface anticyclone should move toward the ________ (direction) and the surface low should move toward the ________ (direction).

   d. On the 300-mb chart, it appears that cold advection is occurring at point_____ and warm advection is occurring at point _____.

2. In Figure 1 on the 500-mb chart:

   a. Draw a contour line and place an "L" on the map to show the position of the upper trough at this level. (Hint: In your textbook, Figure 14.4, p. 359; Fig. 14.10, p. 365; and Fig. 14.12, p. 366, all should help you with this.)

   b. At the 500-mb level, the air is probably slowly rising at point _____ and slowly sinking at point _____.

3. In Figure 1 on the surface weather map:

   a. The storm system is in a stage of development called the ________ ________. (Hint: look at Fig. 14.1, p. 357 in your text.)

   b. Which city is most likely experiencing clearing skies?
      Number _____

   c. Which city is located in the warm sector?
      Number _____

   d. Heavy snow would most likely be falling at which city?
      Number _____

   e. Which cities are most likely experiencing falling air pressure?
      Numbers ________

   (Hint: Figure 14.11, p. 366 in your text, can be of help in answering questions b through e.)

■ ADDITIONAL READINGS

"Bombs and Ultrabombs" by Bud Dorr, *Weatherwise*, Vol. 43, No. 2 (April 1990), p. 76.

"Arctic Hurricanes" by Steven Businger, *American Scientist*, Vol. 79, No. 1 (Jan-Feb 1991), p. 18.

"The Blizzard of '88" by Patrick Hughes, *Weatherwise*, Vol. 40, No. 6 (December 1987), p. 312.

"Profiling Colorado's Christmas Eve Blizzard" by Thomas W. Schlatter, Donald V. Baker and John F. Henz, *Weatherwise*, Vol. 36, No. 2 (April 1983), p. 60.

Skywatch: The Western Weather Guide by Richard A. Keen; Fulcrum Inc., Golden, Colorado, 1987. See especially pp. 12-22.

Storms by William R. Cotton; ASTeR Press, Fort Collins, Colorado, 1990.

Snowstorms Along The Northeastern Coast of the United States: 1955 to 1985 by Paul J. Kocin and Louis W. Uccellini, American Meteorological Society, Boston, MA, 1990.

ANSWERS

MATCHING

1. c
2. a
3. l
4. e
5. j
6. g
7. b
8. k
9. h
10. d
11. f
12. i

FILL IN THE BLANK

1. divergence (of air)
2. baroclinic
3. lee-side lows
4. convergence (of air)
5. jet stream or polar front jet stream
6. west (or left)

MULTIPLE CHOICE

1. a
2. d
3. b
4. c
5. a
6. e

TRUE - FALSE

1. T
2. T
3. F
4. F
5. T
6. F
7. T
8. T
9. T
10. F
11. T

PROBLEMS

1. a. arrows should be parallel to the contour lines
   wind direction at point 1, NW
   wind direction at point 2, SW
   b. convergence at point 1
   divergence at point 2
   c. movement of the surface high should be toward the southeast; movement of the surface low should be toward the northeast
   d. cold advection is occurring at point 1
   warm advection is occurring at point 2
2. a. upper-level low should be to the west of the surface low and a little to the east of the low at the 300-mb level
   b. rising air at point 8
   sinking air at point 7
3. a. open wave
   b. number 6
   c. number 5
   d. number 4
   e. number 3, 4, 5

CHAPTER

15

WEATHER FORECASTING

Chapter Fifteen concentrates on predicting the weather. The beginning of the chapter describes some of the methods and procedures used in making a weather forecast. Here we learn that many ingredients — the current analysis, satellite data, intuition, experience, and the guidance from many computer progs — go into making a weather forecast. The next section looks at some of the problems in predicting the weather. This leads to a discussion that considers some of the effects designed to improve forecasts. This is followed by a section that describes how weather predictions can be made using surface weather maps. The latter part of the chapter examines how a meteorologist makes a prediction with the aid of more sophisticated tools, such as satellite information and prognostic charts.

Some important concepts of this chapter:

1. A weather *watch* indicates that atmospheric conditions are favorable for the development of hazardous weather. A weather *warning* indicates that hazardous weather is either imminent or actually occurring.
2. A persistence forecast is a prediction that future weather will be the same as the present weather. If it is presently raining, a persistence forecast will call for rain.
3. A steady-state forecast (trend method) is a weather prediction based on the assumption that weather systems will continue to move in the same direction and at approximately the same speed as they have been moving.
4. The analogue method of forecasting makes a weather prediction by comparing past weather maps and weather patterns to those of the present.
5. Weather type forecasting categorizes weather patterns into similar groups or "types," using such criteria as the position of the subtropical highs and the upper-level flow.
6. Climatological forecasts are based on the climatology (average weather) of a particular region.
7. Ensemble forecasting is a technique based on running several forecast models (or different versions of a single model), each beginning with slightly different weather information to reflect errors in the measurements. If the different versions agree fairly well, a forecaster can place a high degree of confidence in the forecast. A low degree of confidence means that the models do not agree.
8. For a forecast to show skill, it must be better than a persistence forecast or a climatological forecast.
9. After a number of days, flaws in the computer models and small errors in the data greatly limit the accuracy of weather forecasts.

10. As you move up through the atmosphere, winds that change direction in a clockwise sense (veer with height) indicate that warm air is being brought into the region (warm advection). Winds that change direction in a counterclockwise sense (back with height) indicate that cold air is being brought into the region (cold advection).
11. Surface storms tend to move (1) in a direction parallel to the isobars in the warm sector, and (2) in the direction of greatest pressure fall.

SELF TESTS

MATCH THE FOLLOWING

_____ 1. Forecast chart that shows the atmosphere at some future time

_____ 2. A type of blocking high pressure area

_____ 3. Lines connecting points of equal pressure change

_____ 4. Weather forecast that predicts that the future weather will be the same as the present weather

_____ 5. Designed to replace the AFOS system

_____ 6. These use mathematical equations to describe the atmosphere's behavior

_____ 7. Forecasting method that compares past weather maps and weather patterns to those of the present

_____ 8. This indicates that hazardous weather is imminent or actually occurring

_____ 9. A surface or upper-level chart that shows various weather elements and patterns

_____ 10. Forecasting method that assumes that weather systems will move in the same direction and at the same speed as they have been moving

_____ 11. This indicates that atmospheric conditions are favorable for the development of hazardous weather

_____ 12. A wind that changes direction in a counterclockwise sense is said to be __________.

a. isallobars
b. watch
c. atmospheric model
d. steady state/trend method
e. prog
f. backing
g. AWIPS
h. warning
i. omega high
j. analogue method
k. persistence forecast
l. analysis

■ FILL IN THE BLANK

1. The forecasting of weather using high-speed computers is known as __________ __________ __________.

2. Surface storms tend to move in the region of greatest pressure __________.

3. A weather forecast that is based on the weather conditions averaged over many years is called a __________ __________.

4. An organized band of clouds that look like a comma on a satellite photograph is called a __________ __________.

■ MULTIPLE CHOICE

1. Which of the following displays weather information on a television screen?
   a. WSFO
   b. AFOS
   c. GARP
   d. NEXRAD
   e. WMO

2. Suppose that where you live the weather during July is hot and humid, with afternoon thunderstorms. If for next July 4 you forecast the weather to be "hot and humid with afternoon thunderstorms," you would have made:
   a. a persistence forecast
   b. an ensemble forecast
   c. a climatological forecast
   d. a forecast based on the analogue method
   e. a forecast based on weather types

3. Suppose it is raining and cold outside. You look at the present 500-mb chart and remember a similar weather pattern years ago where the cold rain actually changed into snow. If you use this information to make a forecast of "rain changing to snow," you would have employed which of the following forecasting methods?
   a. steady state/trend
   b. climatological
   c. ensemble
   d. analogue
   e. persistence

4. Normally, the least accurate forecast method of predicting the weather two days into the future is the method called:

   a. persistence
   b. analogue
   c. weather types
   d. numerical weather prediction

5. If a forecaster gives a "degree of confidence" with his or her weather forecast, then the forecaster used a forecasting technique called:

   a. steady state or trend
   b. climatology
   c. persistence
   d. analogue
   e. ensemble

6. Atmospheric models:

   a. are physical models that draw a picture of a developing storm system
   b. are copies of progs that were correctly used in forecasting the weather
   c. use mathematical equations to describe the behavior of the atmosphere
   d. describe atmospheric conditions using clay, blocks, and chart paper
   e. show off the latest attire in meteorological fashions

7. A weather forecast for the immediate future that employs the steady state/trend method is called:

   a. hindcasting
   b. nowcasting
   c. outcasting
   d. guessing

8. Which below is presently a problem with modern-day weather predictions?

   a. there are regions of the world where only sparce observations are available
   b. computer models do not always adequately interpret the surface's influence on the weather
   c. computer-forecast models make assumptions about the atmosphere that are not always correct
   d. the distance between grid points on some models is too large
   e. all of the above

9. Outside, you look up and notice two cloud layers above you. The lower cloud layer is moving from a westerly direction, while the higher cloud layer is moving from a southerly direction. From this observation you conclude that the wind is __________ with height and __________ advection is occurring between the cloud layers.

   a. veering, warm
   b. veering, cold
   c. backing, warm
   d. backing, cold

■ TRUE - FALSE

______ 1. For a weather forecast to show skill it must be better than one based on persistence or climatology.

______ 2. A wind that veers with height, changes direction in a clockwise sense.

______ 3. A probability forecast that calls for a 60 percent chance of rain means that it will rain 60 percent of the time over the forecast area.

______ 4. The greatest improvement in forecasting skill during the past 25 years has been made in forecasting severe storm warnings for hurricanes and tornadoes.

______ 5. An accurate forecast must show skill.

______ 6. Omega highs tend to persist in the same geographic area for many days.

______ 7. AFOS stands for Automation of Field Operations and Services.

______ 8. By examining a surface weather map, the movement of a surface low pressure area can be predicted based upon the orientation of the isobars in the warm sector.

PROBLEMS

Figure 1

1. Suppose it is the end of January and the average weather for your area this time of year during the past 30 years or so has been overcast with fog and drizzle. But, as you can see in Figure 1, outside it is presently cold and snowing. However, a surface weather map indicates that a cold front is approaching from the west and will move through your area in about six hours. On the map you notice that directly behind the front the weather clears and turns much colder. With all of this information, you use the following methods to make a 12-hour weather forecast for your area.

   a. A *persistence forecast* would be:

   b. A *climatological* forecast would be:

   c. A 12-hour weather forecast using the *steady state* or *trend method* would be:

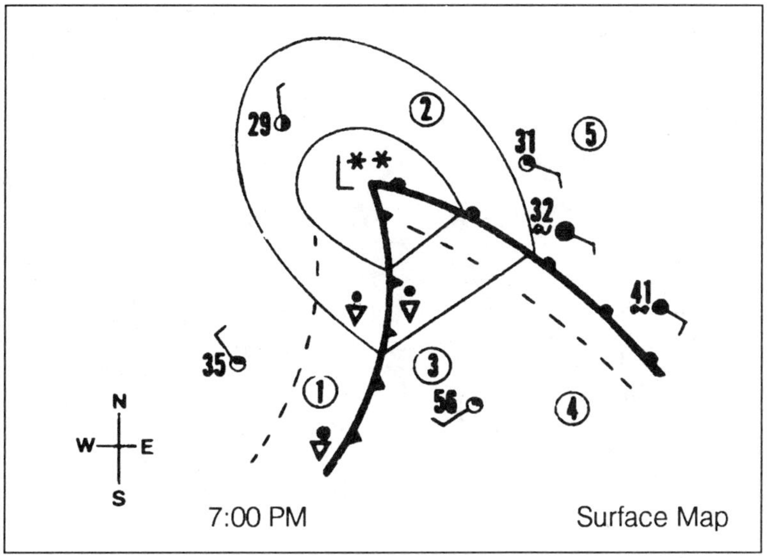

Figure 2

2. The following questions refer to Figure 2, the surface weather map and the midlatitude cyclone. The solid lines on the map are isobars. The dashed lines represent the position of the fronts 12 hours ago. (Symbols and station model information are given in appendix B, pp. 524-525 in your textbook.)

   a. It appears from the orientation of the isobars in the warm sector that the center of the storm will move toward the (give direction) _________.

   b. At which of the 5 positions would you most likely expect to hear the following 12-hour forecasts that run from 7:00 PM to 7:00 AM?

      1. "Partly cloudy continued warm through tonight."
      Forecast for position _____

      2. "Cloudy with the possibility of showers tonight, turning colder by morning."
      Forecast for position _____

      3. "Cloudy and cold tonight with the possibility of light snow beginning by morning."
      Forecast for position _____

      4. "Clearing and colder tonight with continued rising pressures."
      Forecast for position _____

      5. "A gradual increase in cloudiness tonight and continued cold."
      Forecast for position _____

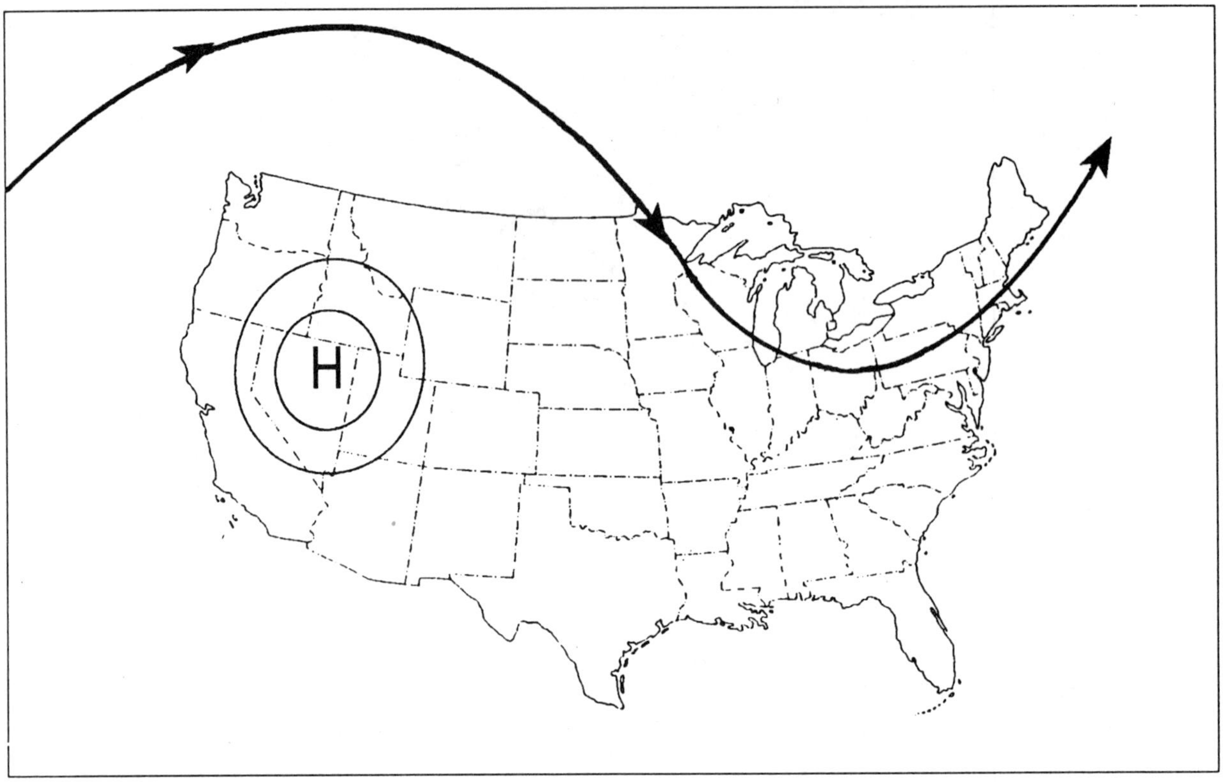

Figure 3

3. Figure 3 illustrates a winter weather pattern. The heavy arrows represent the upper-air flow. The large H (with several isobars) shows the surface position of the Pacific high. If this pattern should persist for a period of several weeks, answer true or false to each regional forecast statement below.

   a. The Pacific northwest will be drier than usual. _____

   b. Extremely cold polar outbreaks will occur over the eastern half of North America. _____

   c. Santa Ana winds will tend to persist over Southern California. _____

   d. The southwestern part of North America will experience extensive cloudiness and heavier than normal precipitation. _____

   e. Chinook winds will occur periodically over the eastern slopes of the southern and central Rocky mountains. _____

   f. The Pacific northwest will be colder than normal. _____

■ ADDITIONAL READINGS

"Numerical Weather Prediction" by Kikuro Miyakoda, *American Scientist*, Vol. 62, No. 5 (Sept-Oct 1974), p. 564.

"Forecasting into Chaos" by Richard Monastersky, *Weatherwise*, Vol. 43, No. 4 (August 1990), p. 202.

"Warm Snowstorms: A Forecaster's Dilemma" by Stanley David Gedzelman and Elaine Lewis, *Weatherwise*, Vol. 43, No. 5 (October 1990), p. 265.

"The Hi-Tech World of T.V. Weathercasting" by Tom Kierein, *Weatherwise*, Vol. 41, No. 2 (June 1988), p. 150.

"Jerome Namias — Pioneering the Science of Forecasting" by Janet Else Basu, *Weatherwise*, Vol. 37, No. 4 (August 1989), p. 190.

"Dreaming of a White Christmas" by David R. Cook, *Weatherwise*, Vol. 39, No. 6 (December 1986), p. 308.

"Forecasting for the Frigid Desert of Antarctica" by Daniel P. Mullen, *Weatherwise*, Vol. 40, No. 6 (December 1987), p. 304.

"A History of Numerical Weather Prediction in the United States" by Philip Duncan Thompson, *Bulletin of the American Meteorological Society*, Vol. 64, No. 7 (July 1983), p. 755.

"The Race to Predict Next Week's Weather" by Richard A. Kerr, *Science*, Vol. 220, No. 4592 (April 1, 1983), p. 39.

"Forecasting the Weather a Bit Better" by Richard A. Kerr, *Science*, Vol. 228, No. 4695 (April 5, 1985), p. 40.

"Predicting the Weather for the Long Term" by Donald Gilman, *Weatherwise*, Vol.36, No. 6 (December 1983), p. 291.

"Computers and the Weather" by J. Murray Mitchell, *Weatherwise*, Vol. 36, No. 5 (October 1983), p. 224.

"A Primer: Using Satellites to Study the Weather" by Vincent Oliver, *Weatherwise*, Vol. 34, No. 4 (August 1981), p. 164.

ANSWERS

MATCHING

1. e	4. k	7. j	10. d
2. i	5. g	8. h	11. b
3. a	6. c	9. l	12. f

FILL IN THE BLANK

1. numerical weather prediction
2. fall (drop)
3. climatological forecast
4. comma cloud

MULTIPLE CHOICE

1. b	4. a	6. c	8. e
2. c	5. e	7. b	9. d
3. d			

TRUE - FALSE

1. T	3. F	5. F	7. T
2. T	4. T	6. T	8. T

PROBLEMS

1. a. persistence forecast: "cold and snow."
   b. climatological forecast: "overcast with fog and drizzle."
   c. steady state/trend method: "snow ending, turning colder with clearing weather."
2. a. storm should move toward the northeast
   b. 1. position 4
      2. position 3
      3. position 2
      4. position 1
      5. position 5
3. a. T
   b. T
   c. T
   d. F
   e. T
   f. F

CHAPTER

16

THUNDERSTORMS AND TORNADOES

Chapter Sixteen examines thunderstorms and tornadoes and the atmospheric conditions that produce them. The beginning of the chapter looks at air-mass thunderstorms and severe thunderstorms. Here we learn that while an air-mass thunderstorm may go through its stages of birth to decay in less than an hour, the supercell thunderstorm, with its updrafts and downdrafts nearly in balance, may exist for hours on end. In the section on severe thunderstorms, the chapter also looks at such phenomena as the gust front, the microburst, and the mesoscale convective complex. The distribution of thunderstorms as well as lightning and thunder are examined next. This is followed by a section that describes how and where tornadoes form and why they are so destructive. The chapter concludes with a discussion of waterspouts.

Some important concepts of this chapter:

1. Air-mass thunderstorms generally form in warm, humid weather, are usually short-lived and rarely produce strong winds and large hail.
2. Severe thunderstorms form in unstable air where the wind speed increases rapidly with height, often in the vicinity of a jet stream. This strong wind shear causes a tilted updraft in the severe thunderstorm's mature stage.
3. Severe thunderstorms (such as the supercell storm and the squall line) are capable of producing large hail (having at least 3/4 inch diameter), strong surface winds (with gusts of 50 knots or more), flash floods, and tornadoes.
4. Strong downdrafts, called microbursts, have been responsible for several airline crashes because microbursts produce rapid changes in wind speed and wind direction — wind shear.
5. While Florida annually experiences the most thunderstorms, the greatest frequency of hailstorms is over the western Great Plains of the United States.
6. Lightning is a visible electrical discharge — a giant spark — that occurs in mature thunderstorms. It may heat the air through which it travels to 30,000°C.
7. Thunder is the sound that results from the rapidly expanding heated air along the channel of the lightning stroke. Contrary to what a once popular song may have said, remember: Thunder only happens when it's lightning.
8. A tornado is a rapidly rotating column of air around an area of intense low pressure that extends from the base of a cumulonimbus cloud.

9. The visible funnel cloud of a tornado is composed of water (cloud) droplets. The funnel cloud is classified as a tornado when the funnel touches the ground.
10. A funnel cloud descends toward the surface as air rushes into its low pressure core, expands, cools, and condenses at successively lower levels beneath the parent thunderstorm.
11. The majority of tornadoes have winds of less than 125 knots, diameters of less than 600 meters, and occur most frequently in the afternoon.
12. When a tornado approaches, don't bother opening windows — seek shelter immediately!
13. On the Fujita scale, of all the tornadoes experienced in the United States annually (the average is over 700), the majority fall into the weak category (F0 and F1) and only a few reach the violent category (F4 and F5). It is the violent tornadoes that account for the majority of tornado-related deaths.
14. Doppler radar has helped scientists gain a great deal of information about tornado-generating thunderstorms.
15. Waterspouts (that form over water) are usually smaller than tornadoes, have weaker winds than tornadoes, and tend to form with developing cumulus clouds.

SELF TESTS

■ MATCH THE FOLLOWING

_____ 1. An enormous thunderstorm that can maintain itself for many hours

_____ 2. The initial stage of an air-mass thunderstorm

_____ 3. The leading edge of a thunderstorm's cold downdraft

_____ 4. A relatively small downburst, less than 4 km wide

_____ 5. Thunderstorms that develop in a line one next to the other, each in a different stage of development

_____ 6. This often looks like a luminous sphere that appears to float in the air or dart for several seconds

_____ 7. An air-mass thunderstorm is most intense during this stage

_____ 8. This is caused by an aircraft flying faster than the speed of sound

_____ 9. Individual thunderstorms that have grown into a large, long-lasting weather system

_____ 10. The rising, spinning column of air inside a severe thunderstorm

_____ 11. Cloud that often forms along the leading edge of a gust front

_____ 12. Downdrafts throughout an air-mass thunderstorm are most likely to occur in this stage

_____ 13. A second surge of electrons that proceeds from the base of a cloud toward the ground

_____ 14. Corona discharge and bluish halo that may appear above pointed objects

a. dissipating

b. sonic boom

c. microburst

d. mesocyclone

e. supercell storm

f. St. Elmo's fire

g. mature

h. multicell thunderstorm

i. dart leader

j. roll cloud

k. cumulus

l. gust front

m. Mesoscale Convective Complex

n. ball lightning

■ ADDITIONAL MATCHING

______	1.	A tornado cloud that does not touch the ground	a.	mesohigh
______	2.	Measures speed at which precipitation is moving horizontally toward or away from you	b.	flash floods
______	3.	Rapidly rotating small whirls that sometimes occur with a large tornado	c.	suction vortices
______	4.	A small area of high pressure created by the cold, heavy air of a thunderstorm's downdraft	d.	Fujita scale
______	5.	A line of thunderstorms that form out ahead of an advancing cold front	e.	wall clouds
______	6.	Classifies tornadoes according to wind speed and damage	f.	funnel cloud
______	7.	A tornado-like feature that forms over water	g.	sferics
______	8.	This marks the boundary where warm, moist air encounters warm, dry air	h.	dry line
______	9.	An area of rotating clouds that extends beneath a severe thunderstorm and from which a funnel cloud may appear	i.	squall line
______	10.	The main cause of wind shear associated with several major airline crashes	j.	microbursts
______	11.	This often results when severe thunderstorms stall or move very slowly	k.	Doppler radar
______	12.	Radio waves produced by lightning	l.	waterspout

■ FILL IN THE BLANK

1. When a tornado is spotted, the National Weather Service issues a __________ __________.

2. A discharge of electricity from or within a cumulonimbus cloud is called __________.

3. The sound produced by rapidly expanding air along the channel of a lightning stroke is called __________.

4. Scattered, isolated, summer thunderstorms that are not severe are called __________ __________.

5. In cloud-to-ground lightning, the stepped leader travels __________ and the return stroke travels __________.

6. On the Fujita scale the strongest, most violent tornado is designated as an F_____.

7. A lighting stroke is seen and 5 seconds later thunder is heard. This means that the lightning stroke is about how many miles away? __________

8. When viewing a severe thunderstorm from the southeast, the most likely place for a tornado to develop is in what section of the thunderstorm? __________

9. Air-mass thunderstorms are most likely to form during what time of the day? __________

10. Lightning associated with thunderstorms that are too far away for the thunder to be heard is referred to as __________ __________.

11. When it appears that tornadoes are likely to form in a particular area, the National Weather Service issues a __________ __________.

■ MULTIPLE CHOICE

1. A funnel cloud is composed mainly of:

   a. ice particles
   b. hail
   c. cloud droplets
   d. raindrops
   e. dust and dirt from the ground

2. When caught in a thunderstorm in an open field, the best thing to do is to:

   a. run for cover under a tree
   b. stand on your hands for as long as you can
   c. immediately lie down flat on the ground
   d. crouch down as low as possible
   e. stand upright with as little surface area on the ground as possible

3. When lightning illuminates the cloud in which it occurs but its flash can not be seen, the lightning is called:

   a. bead lightning
   b. sheet lightning
   c. ribbon lightning
   d. ball lightning
   e. forked lightning

4. In a region where severe thunderstorms with tornadoes are forming, you would *not* expect to observe (Hint: look closely at figures 16.26 and 16.27, p. 431 in your text.):

   a. a strong ridge of high pressure over this region
   b. at 850 mb, a wedge of warm, moist air streaming northward
   c. between 700 mb and 500 mb, a dry tongue of cold air
   d. the polar-front jet stream above the region
   e. an inversion at about the 800 mb level

5. The majority of tornadoes tend to move from:

   a. north to south
   b. northwest to southeast
   c. south to north
   d. southeast to northwest
   e. southwest to northeast

6. The downdraft in an air-mass thunderstorm is created *mainly* by:

   a. evaporating raindrops that make the air cold and heavy
   b. the upper-level wind that dips downward into the thunderstorm
   c. the release of latent heat as ice particles freeze
   d. lightning discharges advancing toward the surface
   e. the melting of snow in the anvil

7. Tornadoes most frequently form in the:

   a. middle of the night
   b. early morning just after sunrise
   c. early evening just after sunset
   d. afternoon
   e. late morning just before noon

8. Severe thunderstorms are different from air mass thunderstorms in that severe thunderstorms are the ones that:

   a. have a strong updraft and downdraft
   b. contain hail
   c. have a tilted updraft in the mature stage
   d. have an anvil
   e. contain thunder and lightning

9. You would most likely expect to see St. Elmo's fire:

   a. over a dry, grassy field
   b. at the top of a tall, dead tree
   c. over a plowed, moist field
   d. over a thick, moist swamp
   e. near the base of a tree

10. The funnel cloud characteristic of a tornado is primarily formed by:

   a. condensation of water vapor that is drawn into the low pressure core of the tornado
   b. dust and dirt picked up from the surface
   c. clouds being funneled by downward air currents coming out of a cumulonimbus cloud
   d. water drawn up from below, into the cloud
   e. smoke from intense lightning activity within the tornado

11. In the United States, the greatest annual frequency of hail occurs in:

   a. Florida
   b. the Mississippi Valley
   c. the western Great Plains
   d. the Pacific Northwest
   e. Texas

12. In the United States, the greatest annual number of thunderstorms occurs in:

   a. Florida
   b. the Mississippi Valley
   c. the Central Plains
   d. the Pacific Northwest
   e. Texas

13. The "Tornado Belt" of the United States is located:

   a. in Florida
   b. in the Middle Atlantic states
   c. in the Ohio Valley
   d. in the Central Plains
   e. along the Gulf Coast

14. For a thunderstorm to spawn a tornado, the updraft in the cloud must:

   a. be stronger than 100 knots
   b. be saturated
   c. be larger than about 20 km
   d. exist all the way to the top of the cloud
   e. rotate

15. The majority of waterspouts:

   a. draw water up into their core
   b. have rotating winds of less than 45 knots
   c. form in an area where winds are descending from a cloud
   d. form with severe thunderstorms
   e. actually form over land

TRUE - FALSE

______ 1. In a severe thunderstorm, hail may actually fall from the base of the anvil.

______ 2. All tornadoes make a distinctive roar.

______ 3. Lightning may occur from one cloud to another.

______ 4. It is now suggested that one *not* open windows as a tornado approaches.

______ 5. A typical diameter of a tornado would be about one mile.

______ 6. In the United States, tornadoes are most frequent during the summer and least frequent during the fall.

______ 7. Thunderstorms that move to the right of the winds aloft tend to cut off the supply of humid air to the thunderstorms north of them.

______ 8. On a Doppler radar screen a tornado vortex signature (TVS) appears as a small region of changing wind speed.

______ 9. Different tornadoes spawned by the same thunderstorm are said to occur in families.

______ 10. All tornadoes rotate counterclockwise.

______ 11. The air behind the leading edge of a gust front is normally warmer than the air ahead of it.

______ 12. Thunder only happens when it's raining.

______ 13. All thunderstorms require rising air.

______ 14. Only Canada experiences more tornadoes than the United States.

______ 15. The winds in a typical tornado are usually less than 125 knots.

______ 16. A hook-shaped echo on a radar screen may indicate the presence of a tornado.

______ 17. The brightest flash of light during lightning is normally caused by the stepped leader.

______ 18. The top part of a thunderstorm usually has a positive charge.

______ 19. A possible reason hailstones become negatively charged and ice crystals positively charged is that in a cloud there is a net transfer of positive ions from warmer objects to colder objects.

______ 20. The "fair-weather" waterspout is normally smaller and less intense than the average tornado.

______ 21. Nighttime thunderstorms over the Central Plains of the United States appear to be related to a low-level southerly jet stream.

______ 22. Lightning can momentarily heat the air to 30,000°C.

______ 23. The planned network of Doppler radars is referred to as NEXDOR.

■ PROBLEMS

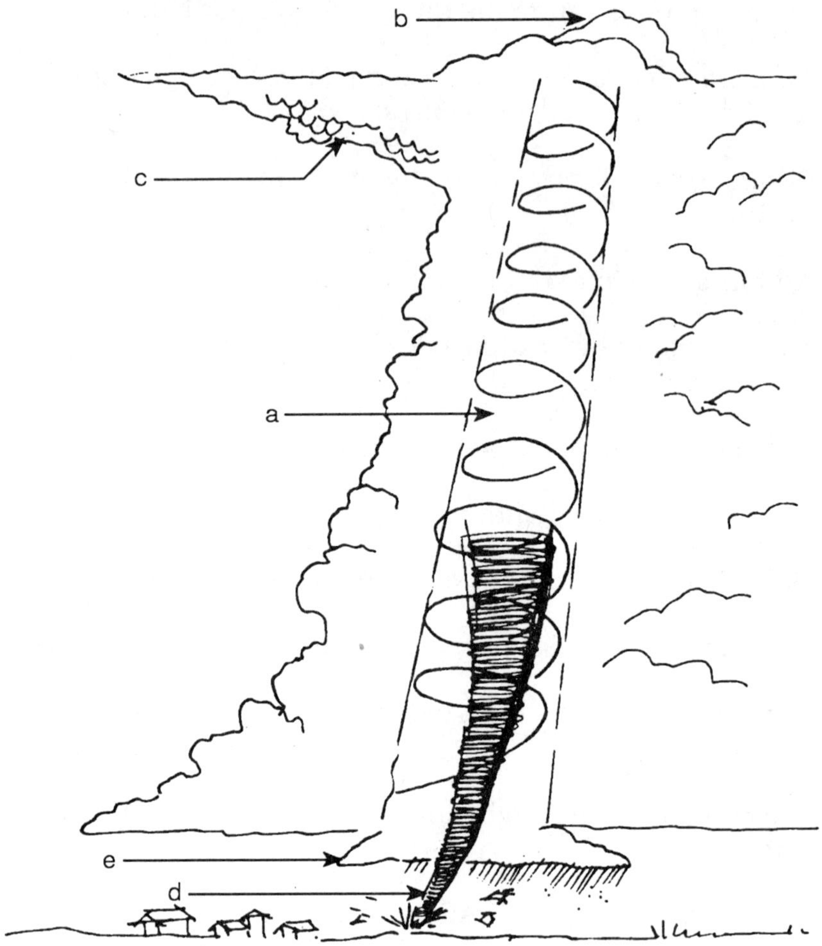

Figure 1

1. Figure 1 represents a side view of a severe thunderstorm and each letter represents a feature of this storm. In the space below, place the name of the feature next to each letter.

   a. ____________________

   b. ____________________

   c. ____________________

   d. ____________________

   e. ____________________

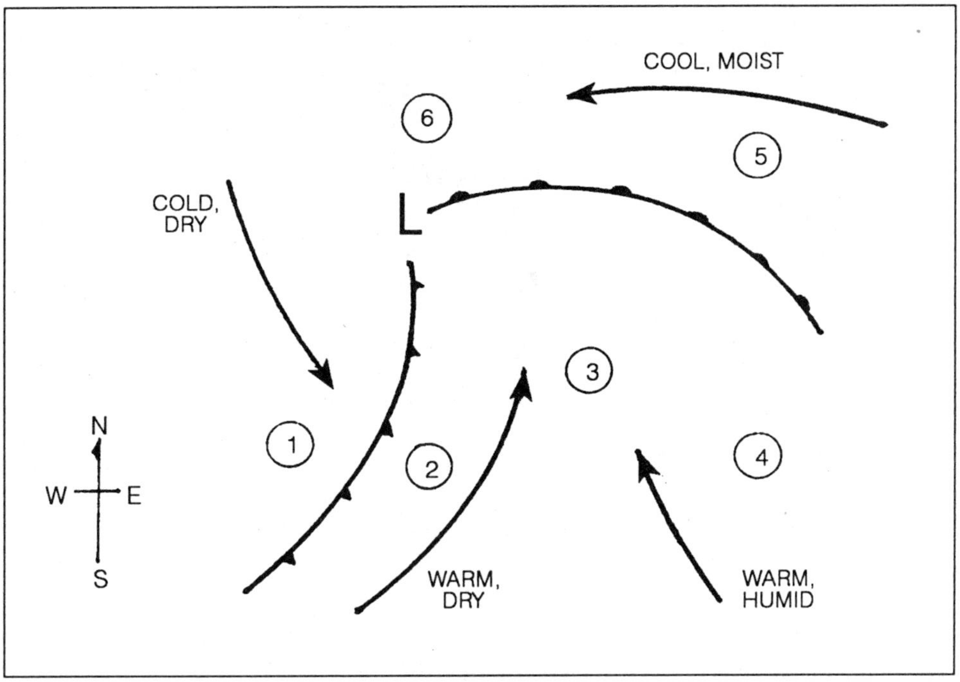

Figure 2

2. Figure 2 represents a springtime surface weather map with a warm front and a cold front associated with an open-wave cyclone. Answer the following questions that pertain to this diagram.

   a. The most likely place for a squall line to develop would be near number _____.

   b. The most likely place for severe thunderstorms to form would be near number _____.

   c. A boxed-off area representing a tornado watch would most likely be placed near number _____.

   d. Number 3 represents a boundary called a __________ __________.

   e. If tornadoes should form near number 3, they would most likely move in a direction toward the __________.

   f. Considering the movement of the tornado in (e), the strongest winds should be on which side of the tornado?
      1. southwest side
      2. northwest side
      3. southeast side
      4. northeast side

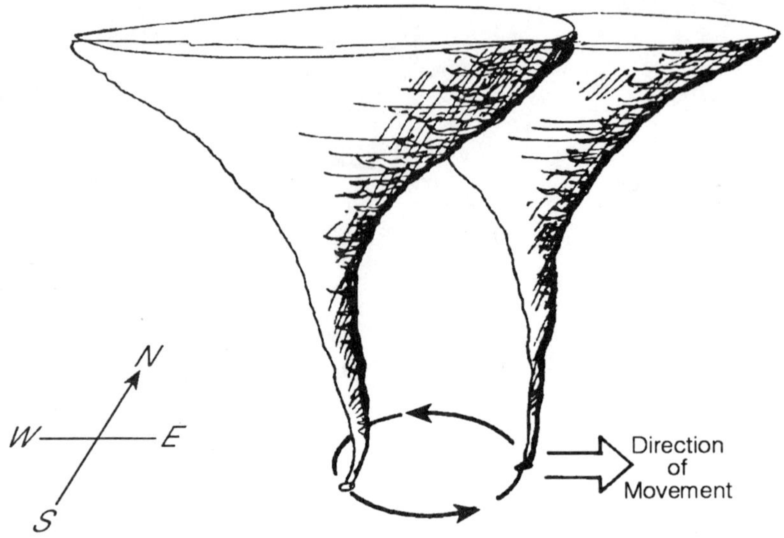

Figure 3

3. Suppose the multi-vortex tornado in Figure 3 is moving toward the east at 40 miles per hour. If the counterclockwise rotational wind speed of the tornado is 130 miles per hour and the counterclockwise rotational wind of the suction vortices are 100 miles per hour, then:

   a. What is the maximum wind speed associated with this multi-vortex tornado? ________

   b. Would the maximum winds occur on the tornado's north, south, east, or west side? ________

   c. In your textbook on p. 430, the Fujita scale (Table 16.1) ranks this tornado as an F_____. It would fall into what category? ________ The expected damage would be ________.

   d. Would you expect the occurrence of a tornado of this magnitude to be common or rare? ________

■ ADDITIONAL READINGS

"The Tornado" by John T. Snow, *Scientific American,* Vol. 250, No. 4 (April 1984), p. 86.

"That's the Way the Building Crumbles" by Hugh Snyder, *Weatherwise,* Vol. 44, No. 3 (June 1991), p. 28.

"The Hesston Tornado: Monster on the Prairie" by Jeff Herzer, *Weatherwise,* Vol. 44, No. 1 (February 1991), p. 23.

"Rugged Rocky Mountain Tornado" by T. Theodore Fujita, *Weatherwise,* Vol. 41, No. 2 (April 1988), p. 80.

"Carolina Tornadoes Kill 59" by Don Witten and Jim Campbell, *Weatherwise,* Vol. 37, No. 3 (June 1984), p. 140.

"The Tri-State Tornado, March 18, 1925" by Jan Brodt, *Weatherwise,* Vol. 39, No. 2 (April 1986), p. 91.

"Tornado Chasing" by Mike Clary, *Weatherwise,* Vol. 39, No. 3 (June 1986), p. 136.

"Inside a Texas Tornado" by Roy S. Hall, *Weatherwise,* Vol. 40, No. 2 (April 1987), p. 72.

"Ten Famous Tornado Outbreaks" by Joseph G. Galway, *Weatherwise,* Vol. 34, No. 3 (June 1981), p. 100.

"Tornado!" by Peter Miller, *National Geographic,* Vol. 171, No. 8 (June 1987), p. 690.

"The Electrification of Thunderstorms" by Earle R. Williams, *Scientific American,* Vol. 259, No. 5 (November 1988), p. 88.

"A New Understanding of Thunderstorms: The Mesoscale Convective Complex" By Robert A. Maddox and J. Michael Fritsch, *Weatherwise,* Vol. 37, No. 3 (June 1984), p. 128.

"The Thunderstorm and its Offspring" by Richard Wood, *Weatherwise,* Vol . 38, No. 3 (June 1985), p. 131.

"Tracking the Dryline" by George W. Bomar, *Weatherwise,* Vol. 38, No. 3 (June 1985) p. 142.

"Leonardo da Vinci and the Downburst" by Stanley David Gedzelman, *Bulletin of the American Meteorological Society,* Vol. 71, No. 5 (May 1990), p. 649.

"Flash Floods" by Richard Addison Wood, *Weatherwise,* Vol. 42, No. 2 (April 1989), p. 93.

"The Microburst: Hazard to Aircraft" by John McCarthy and Robert Serafin, *Weatherwise*, Vol. 37, No. 3 (June 1984), p. 120.

"Lightning!" by Philip Krider and Steven B. Alejandro, *Weatherwise*, Vol. 36, No. 2 (April 1983), p. 71.

"Radar: A Short History" by Stuart Bigler, *Weatherwise*, Vol . 34, No . 4 (August 1981), p. 158.

"NEXRAD — The Coming Revolution in Radar Storm Detection and Warning" by Samuel Milner, *Weatherwise*, Vol. 39, No. 2 (April 1986), p. 72.

ANSWERS

MATCHING

1. e
2. k
3. l
4. c
5. h
6. n
7. g
8. b
9. m
10. d
11. j
12. a
13. i
14. f

ADDITIONAL MATCHING

1. f
2. k
3. c
4. a
5. i
6. d
7. l
8. h
9. e
10. j
11. b
12. g

FILL IN THE BLANK

1. tornado warning
2. lightning
3. thunder
4. air-mass thunderstorms
5. downward, upward
6. F5
7. 1 mile
8. southwest section
9. afternoon
10. heat lightning
11. tornado watch

MULTIPLE CHOICE

1. c
2. d
3. b
4. a
5. e
6. a
7. d
8. c
9. b
10. a
11. c
12. a
13. d
14. e
15. b

TRUE - FALSE

1. T
2. F
3. T
4. T
5. F
6. F
7. T
8. T
9. T
10. F
11. F
12. F
13. T
14. F
15. T
16. T
17. F
18. T
19. T
20. T
21. T
22. T
23. F

PROBLEMS

1. a. mesocyclone
   b. overshooting top
   c. anvil (with mammatus)
   d. tornado
   e. wall cloud
2. a. 3
   b. 3
   c. 3
   d. dry line
   e. northeast
   f. 3, southeast side
3. a. 270 miles per hour
   b. south
   c. F 5, violent, incredible damage
   d. rare occurrence, perhaps only one a year

CHAPTER 17

HURRICANES

Chapter Seventeen is a relatively short chapter that focuses on tropical cyclones. The chapter opens with a section on tropical weather. This is followed by several sections that detail a hurricane's structure, formation, dissipation, and movement. Here we learn that hurricanes are tropical cyclones that are given different names in different regions of the world. In the northern Atlantic and eastern North Pacific they are called hurricanes. In the western North Pacific they are typhoons, and in India and Australia they are tropical cyclones. The chapter concludes by considering how hurricanes are named.

Some important concepts of this chapter:

1. A hurricane is a tropical cyclone, comprised of an organized mass of thunderstorms, with peak winds about a central core (eye) exceeding 64 knots (74 mi/hr).
2. Winds blow counterclockwise about the hurricane's center in the Northern Hemisphere and clockwise about the center in the Southern Hemisphere.
3. Hurricanes form over tropical waters where light winds converge, the humidity is high, and the surface water temperature is typically 79°F or greater.
4. Hurricanes derive their energy from the warm tropical waters and from the latent heat released as water vapor condenses into clouds.
5. In the tropics, hurricanes tend to move from east to west.
6. Hurricanes tend to dissipate rapidly when they move over cold water or over a large land mass.
7. Hurricanes are different from mid-latitude cyclonic storms in that hurricanes have warm central cores, an eye where the air is sinking, more closely spaced isobars and stronger winds, and no distinct fronts.
8. Although the strong winds, huge waves, and high seas of a hurricane can inflict a great deal of damage, it is *flooding* that causes the most destruction.

SELF TESTS

■ MATCH THE FOLLOWING

_____ 1. The zone of intense thunderstorms around the center of a hurricane

_____ 2. A storm of tropical origin that forms over the north Atlantic and eastern north Pacific Ocean

_____ 3. This shows wind-flow patterns on a map

_____ 4. The center of a hurricane

_____ 5. What a hurricane is called in India and Australia

_____ 6. Over the western North Pacific Ocean a storm of tropical origin whose high winds and flooding cause a great deal of destruction

_____ 7. The stage of hurricane development just after the tropical disturbance stage

_____ 8. Weak trough of low pressure in the tropics along which hurricanes occasionally form

_____ 9. This storm has a low-pressure core and weather fronts

_____ 10. The stage of hurricane development just before it becomes a full-blown hurricane

a. streamlines

b. tropical cyclone

c. typhoon

d. midlatitude cyclone

e. eye wall

f. easterly wave

g. hurricane

h. tropical depression

i. tropical storm

j. eye

■ FILL IN THE BLANK

1. Most of the destruction caused by a hurricane is due to ________.

2. An unusual rise in the ocean level along a shore that is due mainly to the winds of a hurricane: ________ ________.

3. The ________ ________ scale relates hurricane central pressure and hurricane winds to possible damage hurricanes can cause.

4. When a hurricane poses a threat to an area, a ________ ________ is issued by the National Weather Service several days before the storm arrives.

■ MULTIPLE CHOICE

1. Which below will *not* cause a hurricane to dissipate?

   a. when it moves over colder water
   b. when it moves over land
   c. when, above the storm, upper-level outflow of air exceeds surface inflow of air

2. The primary source of energy for a hurricane is the:

   a. strong surface winds
   b. heat produced by the sinking of air in the eye
   c. rising of warm air and the sinking of cold air associated with weather fronts
   d. release of latent heat of condensation and warm water
   e. meandering jet stream aloft

3. An area of North America that would *most likely* experience thunderstorms, hurricanes, and tornadoes during the course of a year is the:

   a. region around the Gulf of Mexico
   b. Pacific northwest
   c. south-central Canada
   d. region around the Great Lakes
   e. Great Plains

4. As a hurricane's eye passes directly overhead you would *not* expect to observe:

   a. a rise in temperature
   b. high winds
   c. high clouds
   d. a very low sea level pressure

5. An atmospheric condition that is not conducive to the formation of hurricanes is:

   a. strong upper-level winds
   b. a cold upper-level trough
   c. warm, humid surface air
   d. warm water
   e. a region of converging surface winds

6. Scientists have tried to modify hurricanes by:

   a. igniting huge smoke bombs in the eye of the storm
   b. seeding the hurricane with hair-thin pieces of aluminum
   c. seeding the hurricanes with silver iodide
   d. placing an oil slick over the ocean water and igniting it

7. An Easterly Wave:

   a. moves from west to east
   b. has showers and thunderstorms on its eastern side
   c. has converging winds on its western side

8. Hurricanes do not form:

   a. along the equator
   b. along the ITCZ
   c. with an easterly wave
   d. when the trade wind inversion is weak
   e. when the surface water temperature is warm

■ TRUE - FALSE

______ 1. Hurricanes may contain tornadoes.

______ 2. The vertical structure of a hurricane shows an upper-level outflow of air, and a surface inflow of air.

______ 3. The trade wind inversion is caused by the sinking air along the northward-displaced ITCZ.

______ 4. Intense storms of tropical origin that form over the eastern North Pacific Ocean adjacent to the west coast of Mexico are called hurricanes.

______ 5. Hurricanes only form over water.

______ 6. In the center of a hurricane the surface air pressure is much higher than the air pressure around the periphery of the eye.

_____ 7. A hurricane moving north over the Pacific Ocean adjacent to the west coast of North America will normally survive as a hurricane for a longer time than one moving north over the Atlantic Ocean adjacent to the east coast of North America.

_____ 8. Hurricanes do not form over the South Atlantic Ocean adjacent to South America because of the relatively cold water found there.

_____ 9. Hurricanes in the Northern Hemisphere are similar to middle latitude cyclones in that both generally move from west to east.

_____ 10. The winds in a tropical system known as a tropical storm are greater than the winds in a hurricane.

_____ 11. In the eye of a hurricane, several kilometers above the surface, the air is sinking.

■ PROBLEMS

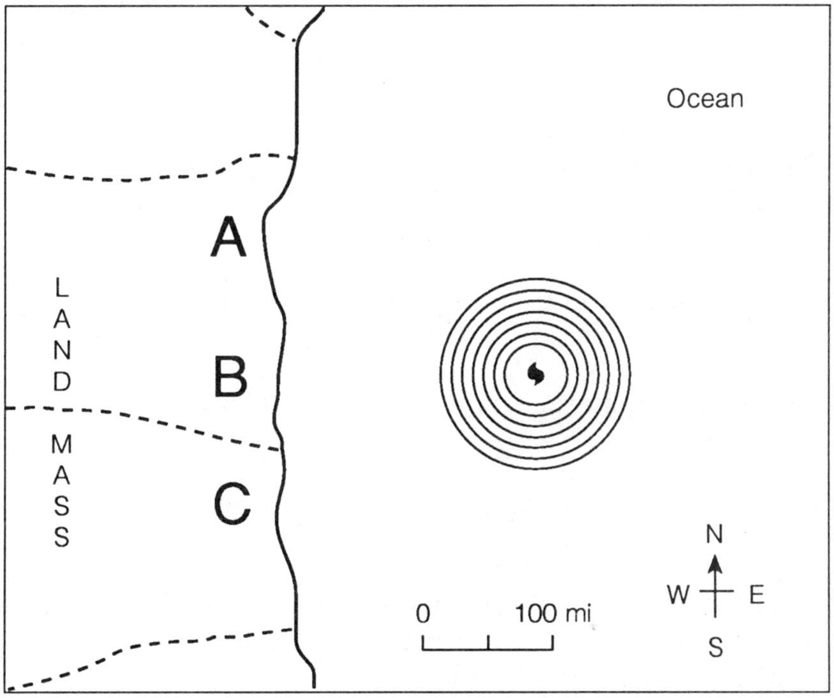

Figure 1
Surface Map

1. Suppose the hurricane in Figure 1 has stalled off the coast of North America. Further suppose that the peak winds around the storm are 125 mi/hr and that its central pressure is 934 mb (27.58 in.). If the hurricane begins to move due westward at 15 mi/hr, answer the following questions assuming that the hurricane experiences no change in its intensity.

   a. On which side of the hurricane (north, south, east or west) would the winds be strongest? __________

   b. How strong would these winds be? __________ mi/hr

   c. On which side of the hurricane (north, south, east, or west) would the winds be weakest? __________

   d. How "weak" would these winds be? __________ mi/hr

   e. Along the coastline, would the strongest storm surge most likely be observed at point A, B, or C? _____

   f. Will the hurricane probably make landfall at point A, B, or C? _____

   g. As the hurricane approaches, from which direction will the wind most likely be blowing at point B? __________

   h. With the maximum winds you calculated in (b), and with the storm's central pressure of 934 mb, on the Saffir-Simpson scale (Table 17.1, p. 454 in your text) determine the category number of this hurricane. __________

   i. You would expect a storm surge of about how many feet with this hurricane? __________ ft

   j. Flooding with this storm should occur for about how many miles inland? __________ mi

   k. As the storm moves onshore, will the lowest sea level pressure probably be observed at point A, B, or C? _____

   l. If this storm is over the North Atlantic Ocean and if it is the twelfth tropical storm or hurricane named in this region during the 1992 hurricane season, what would be the name of this storm? __________ (Hint: Look at Table 17.2, p. 455 in your textbook.)

ADDITIONAL READINGS

"Toward a General Theory of Hurricanes" by Kerry A. Emanuel, *American Scientist*, Vol. 76, No. 4 (July-Aug 1988), p. 36.

"The Great Galveston Hurricane" by Patrick Hughes, *Weatherwise*, Vol. 43, No. 4 (August 1990), p. 190.

"The Siege of New England" by Hugh Cobb, *Weatherwise*, Vol. 42, No. 5 (October 1989), p. 262.

"The Great Hurricane of 1938" by David M. Ludlum, *Weatherwise*, Vol. 41, No. 4 (August 1988), p. 214.

"The Impossible Hurricane — Could it Happen Again?" by A. James Wagner, *Weatherwise*, Vol. 41, No. 5 (October 1988), p. 279.

"The Ultimate Storm" by Bruce Cotton, *Weatherwise*, Vol. 38, No. 5 (October 1985), p. 248.

"The Chesapeake-Potomac Hurricane of 1933" by Hugh D. Cobb, III, *Weatherwise*, Vol.44, No. 4 (Aug-Sept 1991), p. 24.

"Flying into the Eye of a Hurricane" by Jeffrey M. Masters, *Weatherwise*, Vol. 40, No. 3 (June 1987), p. 128.

"Hurricanes Haunt Our History" by Patrick Hughes, *Weatherwise*, Vol. 40, No. 3 (June 1987), p. 134.

"Tropical Cyclones of the Australian Region" by P. Price, *Weatherwise*, Vol. 36, No. 3 (June 1983), p. 118.

ANSWERS

MATCHING

1. e
2. g
3. a
4. j
5. b
6. c
7. h
8. f
9. d
10. i

FILL IN THE BLANK

1. flooding
2. storm surge
3. Saffir-Simpson
4. hurricane watch

MULTIPLE CHOICE

1. c
2. d
3. a
4. b
5. a
6. c
7. b
8. a

TRUE - FALSE

1. T
2. T
3. F
4. T
5. T
6. F
7. F
8. T
9. F
10. F
11. T

PROBLEMS

1. a. north
   b. 140 mi/hr
   c. south
   d. 110 mi/hr
   e. point A
   f. point B
   g. from the north and northeast
   h. category 4
   i. between 13 ft and 18 ft
   j. about 6 mi
   k. point B
   l. Lisa

CHAPTER

18

GLOBAL CLIMATE

Chapter Eighteen examines the large scale aspects of climate. The chapter begins by looking at the global patterns of temperature and precipitation. This is followed by several sections that detail the world's climatic regions. Here we learn that the Köppen scheme of classifying climate employs five major climatic types, each being designated by a capital letter. Tropical moist climates are A, while dry climates are B. Mid-latitude moist climates with mild winters are designated as C, and those with severe winters as D. Polar climates are given the letter E. Each group contains subregions that describe specific regional characteristics, such as seasonal changes in temperature and precipitation. The chapter concludes by examining highland (H) climates — those found in mountainous areas, where rapid changes in elevation bring about sharp changes in climatic zones over a relatively short distance.

Some important concepts of this chapter:

1. Climate is the accumulation of daily and seasonal weather events over a long period of time, as well as the extremes of weather within a specified area.
2. The climatic controls are the factors that govern the climate of any given region. (The seven major climate controls are found on p. 460 in your textbook.)
3. The hottest places on earth tend to occur in the subtropical deserts of the Northern Hemisphere, where clear skies and sinking air coupled with low humidity, and a high summer sun beating down upon a relatively barren landscape produce extreme heat.
4. The coldest places on Earth tend to occur in the interior of high-latitude landmasses. The Antarctic holds the record for extreme cold.
5. The rainiest places in the world are located on the windward side of mountains where warm, humid air rises upslope.
6. In the middle of a mid-latitude continent, summers are usually wetter than winters.
7. Tropical moist climates are found in low latitudes where abundant rainfall exists, the noon sun is always high, day and night are of nearly equal length, every month is warm, and no real winter season exists.
8. Dry climates prevail where potential evaporation exceeds precipitation. Subtropical deserts, like the Sahara, are mainly the result of sinking air associated with a subtropical high. Other deserts form on the leeward side of mountains due to the rainshadow effect. Some deserts form in response to both of these effects.

9. Coastal deserts can experience considerable low cloudiness, fog, and even drizzle.
10. Mid-latitude climates are characterized by distinct winter and summer seasons. Lower latitudes tend to have longer, hotter summers and shorter, milder winters than do higher latitudes. Summers become shorter and winters longer and colder as one moves poleward.
11. Polar climates prevail at high latitudes where winters are severe and there is no real summer season.

SELF TESTS

MATCH THE FOLLOWING

_____ 1. These plants are capable of surviving a prolonged period of dryness

_____ 2. The majority of the southeastern section of the United States has this type of climate

_____ 3. Name given to the small climatic region near the ground

_____ 4. A grass-covered treeless plain that has a semiarid climate

_____ 5. One would expect to observe savana grass in this climate

_____ 6. This is considered a climatic control

_____ 7. This type of vegetation grows where the average temperature of the warmest month is above freezing but below 10°C (50°F)

_____ 8. The freezing of ground to great depths

_____ 9. One would expect to observe this type of vegetation in a subpolar climate

_____ 10. Abundant yearly rainfall and high temperatures of this climatic type combine to produce a dense, broadleaf, evergreen forest

a. steppe
b. tropical wet
c. ocean currents
d. tundra
e. microclimate
f. tropical wet-and-dry
g. humid subtropical
h. permafrost
i. xerophytes
j. taiga

FILL IN THE BLANKS

1. This climate exists where potential evaporation exceeds precipitation: __________.

2. Another name used to describe a dry-summer subtropical climate: __________.

3. The Köppen scheme for classifying climates employs annual and monthly averages of __________ and __________.

4. This climatic type experiences no distinct summer season: __________ __________.

5. List the seven major climatic controls:

   a. ____________________

   b. ____________________

   c. ____________________

   d. ____________________

   e. ____________________

   f. ____________________

   g. ____________________

6. This climatic type experiences no distinct winter season: __________ __________.

MULTIPLE CHOICE

1. A climatic type that experiences a greater temperature variation between day and night than between the warmest and coldest months is:

   a. subpolar
   b. arid
   c. tropical wet
   d. humid subtropical
   e. dry-summer subtropical

2. This climate type covers more land mass than any other:

   a. tropical climates, A
   b. dry climates, B
   c. moist subtropical mid-latitude climates, C
   d. moist continental climates, D
   e. polar climates, E

3. One would expect to observe the melting of an upper layer of permafrost in this climatic type:

   a. polar tundra
   b. polar ice cap
   c. humid continental with warm summers
   d. marine
   e. humid subtropical

4. A city that would most likely experience a dry summer is:

   a. Baltimore, Maryland
   b. Chicago, Illinois
   c. San Francisco, California

5. This region of the world experiences the lowest average temperatures:

   a. the Arctic
   b. Northern Siberia
   c. Northern Canada
   d. Alaska
   e. the Antarctic

6. This climatic type experiences cool, wet winters and mild to hot, dry summers:

   a. humid continental climates
   b. tropical wet-and-dry climates
   c. humid subtropical climates
   d. Mediterranean climates
   e. tropical monsoon climates

7. The highest temperatures in the world are normally experienced with this climatic type:

   a. tropical moist climates
   b. dry climates
   c. humid subtropical climates
   d. humid continental climates
   e. highland climates

8. Moist continental (D) climates are not observed in:

   a. Canada
   b. Alaska
   c. Europe
   d. the northern plains of the United States
   e. South America

9. Of the climatic types listed below, which would normally experience the greatest annual range in temperature?

   a. subpolar climate
   b. humid subtropical climate
   c. polar ice cap climate
   d. arid climates
   e. tropical wet-and-dry climates

10. The *primary* reason for the dry summer subtropical climate in North America is due to the position of the:

   a. ITCZ
   b. polar front
   c. subtropical high

11. The two primary weather features that influence the tropical wet-and-dry climate are the:

   a. polar front and ITCZ
   b. subtropical highs and ITCZ
   c. subtropical highs and polar fronts
   d. subtropical highs and subpolar lows
   e. subpolar lows and ITCZ

12. This climatic type has relatively mild winters and hot, humid summers:

   a. humid continental climate
   b. tropical wet-and-dry climate
   c. humid subtropical climate
   d. semi-arid climate
   e. marine climate

13. Short bunch grass, scattered low bushes, trees, or sagebrush categorizes a region known as:

   a. tundra
   b. taiga
   c. savanna
   d. steppe

14. This climatic type experiences adequate precipitation, warm summers, and very cold winters with snowstorms and blustery winds:

    a. humid subtropical
    b. humid continental
    c. semi arid
    d. polar tundra
    e. polar ice cap

■ TRUE - FALSE

______ 1. One would expect to observe a tropical rain forest in a tropical wet-and-dry climate.

______ 2. Rainshadow deserts are normally observed on the downwind side of a mountain.

______ 3. The Köppen scheme for classifying climates does not use the letter H.

______ 4. The rainiest places in the world are usually located on the windward side of mountains.

______ 5. Afternoon temperatures in a tropical wet climate are normally much higher than summer afternoon temperatures in the middle latitudes.

______ 6. In California, one could experience all of Köppen's climatic types.

______ 7. Deserts that experience low clouds and drizzle tend to be found mainly on the western side of continents.

______ 8. In a tropical wet-and-dry climate, the dry season occurs in summer, or during the high sun period.

______ 9. The climate of an area about the size of a city would be described as mesoclimate.

______ 10. Köppen classified a polar ice cap climate as one where the average temperature of the warmest month averages 32°F or below.

______ 11. According to Köppen, the climate of a city positioned high in the mountains would be classified as highland.

______ 12. Most of Canada has a climate classified as polar.

______ 13. Another name for the taiga climate is boreal climate.

______ 14. The semi-arid climate marks the transition between the arid and humid climatic regions.

______ 15. According to Köppen, the driest of all climates would be classified as BW.

■ PROBLEMS

1. On the map of the Northern Hemisphere (Figure 1), draw the approximate boundaries for Köppen's five major climate regions listed below:

   a. Tropical Moist Climates (Group A)
   b. Dry Climates (Group B)
   c. Moist Subtropical Mid-latitude Climates (Group C)
   d. Moist Continental Climates (Group D)
   e. Polar Climates (Group E)

Figure 1

2. a. Table 1 contains the mean annual precipitation and mean annual temperature records for five cities. Based on Köppen's climatic types, determine the climate of each city. (Hint: Table 18.1, p. 469, can be of assistance.)

Table 1

		Jan	Feb	Mar	April	May	June	July	Aug	Sept	Oct	Nov	Dec	Year
CITY 5	temp. (°F)	13	17	26	40	50	56	62	60	51	42	28	19	39
	precip. (in.)	0.5	0.5	0.8	1.0	2.3	3.1	2.5	2.3	1.5	0.7	0.7	0.6	16.5
CITY 34	temp. (°F)	78	80	82	83	83	81	80	80	80	80	79	78	80
	precip. (in.)	0.2	0.0	0.2	0.2	3.0	11.7	5.3	5.1	7.2	9.6	2.3	0.2	45.0
CITY 26	temp. (°F)	32	35	43	55	64	74	78	77	70	58	44	35	55
	precip. (in.)	2.0	2.0	3.1	3.7	3.7	4.3	3.3	3.0	2.8	2.9	2.6	2.0	35.4
CITY 6	temp. (°F)	-26	-26	-13	3	21	37	46	46	37	21	0	-15	11
	precip. (in.)	0.3	0.4	0.4	0.7	0.5	1.0	1.8	1.7	1.5	1.2	0.8	0.6	10.9
CITY 24	temp. (°F)	31	36	41	48	55	62	70	67	60	51	40	33	50
	precip. (in.)	1.0	1.0	0.7	0.5	0.5	0.4	0.2	0.2	0.2	0.6	0.6	0.9	6.8

Climate for City 5 ____________________

Climate for City 34 ____________________

Climate for City 26 ____________________

Climate for City 6 ____________________

Climate for City 24 ____________________

b. The number for each city corresponds to the number found on the map on p. 532 in your textbook. Place each city in its proper position on your map (Figure 1). Then check the climate of each city you determined with those given in your textbook in Table F, beginning on p. 533.

ADDITIONAL READINGS

"Temperature Extremes: A New Cold Champ" by David H. Hickcox, *Weatherwise*, Vol. 44, No. 1 (February 1991), p. 48.

"A December to Remember" by Richard Heim, Kenneth Dewey, Mark Anderson and Daniel Leathers, *Weatherwise,* Vol. 43, No. 6 (December 1990), p. 329.

"The Weather Where You Live" by David A. Robinson and David M. Ludlum, *Weatherwise*, Vol. 42, No . 6 (December 1989), p. 328.

"How Hot Can it Get?" by David Hickcox, *Weatherwise*, Vol. 41, No. 2 (June 1988), p. 157.

"Snowiest Major Cities" by Peter R. Chaston, *Weatherwise*, Vol . 39, No. 1 (February 1986), p. 44.

"How Often Does it Rain Where You Live?" by Nolan J. Doesken and William P. Eckrich, *Weatherwise*, Vol. 40, No. 4 (July-Aug 1987), p. 200.

"Climate and the Timberline in the Appalachians" by Robert J . Leffler, *Weatherwise*, Vol. 34, No. 3 (June 1981), p. 116.

"The Climates of Wyoming" by Brooks E. Martner, *Weatherwise*, Vol. 34, No. 5 (October 1981), p. 204.

"The Weather and Climate of Alaska" by Sue Ann Bowling, *Weatherwise*, Vol. 33, No. 5 (October 1980), p. 197.

"Statistical Descriptors of Climate" by Nathaniel B. Guttman, *Bulletin of the American Meteorological Society*, Vol. 70, No. 6 (June 1989), p. 602.

ANSWERS

MATCHING

1. i
2. g
3. e
4. a
5. f
6. c
7. d
8. h
9. j
10. b

FILL IN THE BLANKS

1. dry climate
2. Mediterranean climate
3. temperature and precipitation
4. polar climates
5. a. intensity of sunshine and its variation with latitude
   b. distribution of land and water
   c. ocean currents
   d. prevailing winds
   e. positions of high-and-low pressure areas
   f. mountain barriers
   g. altitude
6. tropical climates

MULTIPLE CHOICE

1. c
2. b
3. a
4. c
5. e
6. d
7. b
8. e
9. a
10. c
11. b
12. c
13. d
14. b

TRUE - FALSE

1. F
2. T
3. F
4. T
5. F
6. F
7. T
8. F
9. T
10. T
11. T
12. F
13. T
14. T
15. T

PROBLEMS

1. Check your climatic boundaries on your map (Figure 1), with Figure 18.6, on p. 472 in your textbook.
2. City 5, (Calgary, Canada) Climate Dfb
   City 34, (Managua, Nicaragua) Climate Aw
   City 26, (St. Louis, MO) Climate Cfa
   City 6, (Chesterfield, Canada) Climate ET
   City 24, (Reno, NV) Climate BSk

CHAPTER

19

CLIMATE MODIFICATION AND CHANGE

Chapter Nineteen considers some of the ways climate can be modified and changed. The chapter begins by looking at climate modification of the urban environment. Next it investigates the changing climate on a global scale. The chapter concludes with a discussion of the many theories that attempt to explain why the climate has varied over the years and how the climate may change in the future. At this point we learn that the riddle of the earth's changing climate is not completely understood, because a change in one variable in the complex climate system almost always changes other variables.

Some important concepts of this chapter:

1. On the average, cities tend to be warmer and more polluted than rural areas.
2. Severe air pollution episodes usually occur where a strong anticyclone and an upper-level ridge become entrenched over an area.
3. Emissions of sulfur and nitrogen oxides can be transformed into acids that fall to the surface as acid precipitation.
4. The earth's climate is constantly undergoing change; studies suggest that the global climate throughout much of the geologic past was much warmer than it is today.
5. The shifting of continents along with volcanic activity and mountain building described by the theory of plate tectonics proposes how climatic variation can take place over millions of years.
6. The Milankovitch theory — in association with other natural factors — proposes that small variations in the earth's orbit are responsible for variations in the amount of sunlight reaching the earth and, hence, for glacial advances and retreats during the past 2 million years.
7. Volcanic eruptions, rich in sulfur, may be responsible for cooler periods in the geologic past.
8. Fluctuations in solar output (brightness) may account for climatic changes over time scales of decades and centuries.
9. Many numerical climate models predict that doubling the present concentration of CO_2 (which is about 350 ppm) will enhance the atmospheric greenhouse effect and warm the surface air between 3°F and 9°F during the next century. The models also

predict, however, that for this amount of warming to occur, the atmospheric concentration of *water vapor* must also increase.

10. At this point, it is not certain how clouds and oceans will respond to a warmer earth. Recent studies show, however, that overall clouds tend to cool the climate system. A world without clouds would be a warmer world.

11. Several studies indicate that the earth's surface has warmed by about 0.5°C (about 1°F) over the past 100 years, with the decade of the 1980s being the warmest of this century so far.

12. Although no one can unequivocally demonstrate that the recent warming trend is due to increasing levels of CO_2 and other greenhouse gases (CH_4, N_2O, and CFCs), many (but not all) scientists suspect that these greenhouse gases are responsible for the warming.

13. In certain regions of the world, overgrazing and deforestation are changing the reflectivity of the surface and, in some cases, rendering the land useless.

SELF TESTS

MATCH THE FOLLOWING

_____ 1. This refers to a time when there were few sunspots

_____ 2. The study of annual growth rings of trees

_____ 3. At night, this blows into a city from the surrounding rural areas

_____ 4. An increase in the desert conditions of a region

_____ 5. When an initial process is reinforced by another process it is referred to as this

_____ 6. The cooling trend that occurred from about 1550 to about 1850

_____ 7. The most potent greenhouse gas

_____ 8. Another name for the Pleistocene epoch

_____ 9. This greenhouse gas has doubled in concentration over the past 100 years or so

_____ 10. The dark, cold, and gloomy conditions that presumably would be brought on by nuclear war

_____ 11. Another name for the year 1816

_____ 12. A pattern of reversing stratospheric winds above the tropics

a. country breeze

b. water vapor

c. positive feedback mechanism

d. nuclear winter

e. Maunder Minimum

f. quasi-biennial oscillation

g. Ice Age

h. desertification

i. year without a summer

j. Little Ice Age

k. dendrochronology

l. carbon dioxide

FILL IN THE BLANKS

1. Volcanoes that have the greatest impact on global climate appear to be those rich in this gas: __________.

2. Climate models predict that in order for increasing levels of CO_2 to raise air temperatures between 3°F and 9°F, this gas must also increase in concentration: __________.

3. Overall, clouds have a net __________ effect on climate.

4. If the earth were in a warming trend, increasing the amount of water vapor in the atmosphere (but not clouds) would most likely produce a __________ feedback mechanism.

5. A theory that suggests that variations in the earth's orbit are responsible for the advance and retreat of ice over periods of 10,000 to 100,000 years is the __________ __________.

6. The higher air temperatures of cities contrasted to the cooler surrounding rural areas is referred to as the __________ __________ __________.

MULTIPLE CHOICE

1. Normally, air pollution episodes are most severe when:
   a. the country breeze is weak
   b. a storm system is developing west of the area
   c. a cold upper-level low moves over the area
   d. a large, deep anticyclone stagnates over the area
   e. a warm front passes through the area

2. A negative feedback mechanism:
   a. weakens an initial change in an atmospheric process
   b. reinforces an initial change in an atmospheric process
   c. either reinforces or weakens an initial change in an atmospheric process

3. Most climate models predict that a gradual increase in global CO_2 over the next 100 years will most likely bring about:
   a. a decrease in surface air temperature
   b. an increase in surface air temperature
   c. no change in surface air temperature

4. The theory that explains how glacial material can be observed today near sea level at the equator, even though sea level glaciers probably never existed at the equator, is the:

   a. volcanic dust theory
   b. Maunder theory
   c. theory of plate tectonics
   d. Milankovitch theory
   e. changing solar output theory

5. During a period when the earth's orbital tilt is at a maximum, which would probably be true?

   a. There would be less seasonal variation between summer and winter.
   b. There would be a lesser likelihood of glaciers in high latitudes.
   c. In polar regions, more snow would probably fall during the winter.

6. Of all the conditions listed below, which would *most likely* produce warming of the air at the earth's surface?

   a. increase the amount of sulfur-rich particles in the atmosphere
   b. decrease the energy output of the sun
   c. increase the amount of snow cover over the face of the globe
   d. increase the amount of low-level global cloudiness
   e. increase the amount of high-level global cloudiness

7. Downwind of a large industrial area, which of the following is most likely *not* to occur?

   a. higher air temperatures
   b. an increase in average annual precipitation
   c. rain that is more acidic

8. The wobble of the earth on its axis refers to:

   a. obliquity of the earth's axis of rotation
   b. eccentricity of the earth's orbit
   c. precession of the equinox

9. If the earth were in a cooling trend, which process below would *most likely* act as a *positive* feedback mechanism?

   a. decreasing the snow cover around the earth
   b. increasing the amount of cloud cover around the earth
   c. increasing the water vapor content of the air

10. Which below is *not* one of the possible consequences of global warming predicted by climate models?

    a. a reduction in average precipitation over certain areas
    b. a cooling of the upper atmosphere
    c. a drop in the rate of ozone destruction
    d. accumulations of additional snow in Antarctica
    e. a lowering of sea level

11. Many scientists feel that the recent warming trend experienced over the Northern Hemisphere could be the result of:

    a. increasing levels of CO_2 and other greenhouse gases
    b. fewer snowfalls and hence a lower surface albedo
    c. light colored particles in the stratosphere
    d. a decrease in energy emitted from the sun
    e. increasing volcanic eruptions

■ TRUE - FALSE

______ 1. As the number of sunspots increases, it appears that the sun's energy output also increases.

______ 2. Several studies suggest that during the past century global temperatures have increased slightly.

______ 3. On clear, cold winter nights, cities tend to cool more quickly than rural areas, and have lower temperatures.

______ 4. It appears that throughout much of the earth's history, the climate was much cooler than it is today.

______ 5. Most of the Sahel in North Africa is a sand-covered desert.

______ 6. Studies reveal that during colder glacial periods, levels of carbon dioxide were lower than during warmer interglacial periods.

______ 7. The Milankovitch cycles, in association with other natural factors, explain how glaciers may advance and retreat over periods of ten thousand years to one hundred thousand years.

______ 8. Large volcanic eruptions, with an ash veil that enters the stratosphere, tend to warm the climate at the surface.

______ 9. Sunshine in a city is usually less than that in the surrounding rural areas because of the pollution in the city.

______ 10. Atmospheric stagnation conditions are normally brought on by slow moving anticyclones.

______ 11. The CLIMAP project found strong evidence that climatic variations during the past hundred thousand years were closely associated with variations in the sun's energy output.

______ 12. The higher the ratio of oxygen 18 to oxygen 16 in the shells of organisms that lived in the sea during the geologic past, the colder the climate at that time.

______ 13. Most climate models predict that if average global temperatures rise by about 5°F or more, average global precipitation will increase.

______ 14. A strong country breeze would probably be associated with a strong heat island.

______ 15. Dry deposition is another name for acid rain.

______ 16. Aerosols appear to have the effect of only increasing surface temperature in urban areas.

______ 17. The climatic optimum occurred shortly after the Little Ice Age.

______ 18. The higher the concentration of oxygen 16 in the ice caves of Antarctica and Greenland, the colder the air at that time.

______ 19. If global temperatures continue to rise, the oceans may either increase or reduce the amount of CO_2 in the atmosphere.

______ 20. It is not certain at this time how clouds and oceans will respond to a warmer earth.

______ 21. The biogeophysical feedback mechanism in the Sahel relates an increase in vegetation to an increase in surface temperatures and a reduction in rainfall.

______ 22. Extensive glaciation is more likely during periods of cooler summers and milder winters.

■ PROBLEMS

1. a. In Figure 1, the mixing depth is about how many feet thick? __________ ft

   b. In Figure 1, the largest concentration of pollutants would probably be observed in which layer? __________

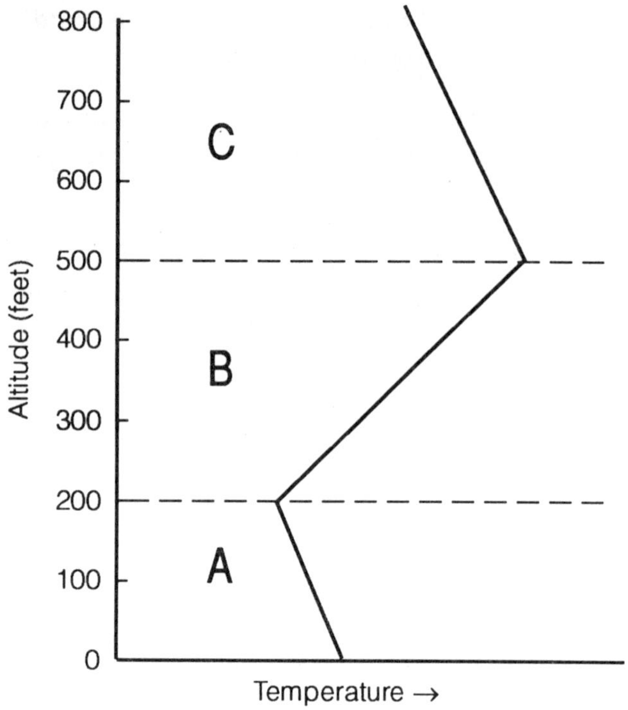

Figure 1

■ ADDITIONAL READINGS

"El Niño and Variations in Climate" by Eugene M. Rasmusson, *American Scientist*, Vol, 73, No. 2 (March-April 1985), p. 168.

"Climate Modeling" by Stephen H. Schneider, *Scientific American*, Vol. 256, No. 5 (May 1987), p. 72.

"Drought in Africa" by Michael H. Glantz, *Scientific American*, Vol.256, No. 6 (June 1987), p. 34.

"The Challenge of Acid Rain" by Volker A. Mohnen, *Scientific American*, Vol. 259, No. 2 (August 1988), p. 30.

"Global Climatic Change" by Richard A. Houghton and George M. Woodwell, *Scientific American*, Vol. 260, No. 4 (April 1989), p. 36.

"The Changing Climate" by Stephen H. Schneider, *Scientific American*, Vol. 261, No. 3 (September 1989), p. 70.

"The Great Climate Debate" by Robert M. White, *Scientific American*, Vol, 263, No. 1 (July 1990), p. 36.

"Volcanic Activity and Climatic Changes" by R. A. Bryson and B. M. Goodman, *Science*, Vol. 207, No. 4435 (March 1980), p. 1041.

"Global Warming Trends" by Philip D. Jones and Tom M. L. Wigley, *Scientific American*, Vol. 263, No. 2 (August 1990), p. 84.

"Carbon Dioxide and Future Climate" by J. Murray Mitchell, Jr., *Weatherwise*, Vol. 44, No. 4 (Aug-Sept 1991), p. 17.

"A Second Look at the Impacts of Climate Change" by Jesse H. Ausubel, *American Scientist*, Vol. 79, No. 3 (May-June 1991), p. 210.

"Plateau Uplift and Climatic Change" by William F. Ruddiman and John E. Kutzbach, *Scientific American*, Vol. 264, No. 3 (March 1991), p. 66.

"Frozen Earth: Explaining the Ice Ages" by R. V. Fodor, *Weatherwise*, Vol. 35, No. 3 (June 1982), p. 108.

"1816 — The Year Without a Summer" by Patrick Hughes, *Weatherwise*, Vol. 32, No. 3 (June 1979), p. 108.

"The Aerial Fertilization Effect of CO_2 and its Implications for Global Carbon Cycling and Maximum Greenhouse Warming" by Sherwood B. Idso, Bulletin of the American Meteorological Society, Vol. 72, No. 7 (July 1991), p. 962.

"Is Our World Warming?" by Samuel W. Matthews, *National Geographic*, Vol. 178, No. 4 (October 1990), p. 66.

ANSWERS

MATCHING

1. e	4. h	7. b	10. d
2. k	5. c	8. g	11. i
3. a	6. j	9. l	12. f

FILL IN THE BLANKS

1. sulfur
2. water vapor
3. cooling
4. positive
5. Milankovitch theory
6. urban heat island

MULTIPLE CHOICE

1. d	4. c	7. a	10. e
2. a	5. b	8. c	11. a
3. b	6. e	9. b	

TRUE - FALSE

1. T	7. T	13. T	19. T
2. T	8. F	14. T	20. T
3. F	9. T	15. F	21. F
4. F	10. T	16. F	22. T
5. F	11. F	17. F	
6. T	12. T	18. T	

PROBLEMS

1. a. 200 feet
   b. layer A